INTRODUCTION TO MATLAB® & SIMULINK

A Project Approach

Third Edition

INTRODUCTION TO MATLAB® & SIMULINK

A Project Approach

Third Edition

O. BEUCHER
and
M. WEEKS

INFINITY SCIENCE PRESS LLC
Hingham, Massachusetts
New Delhi

INFINITY SCIENCE PRESS LLC
11 Leavitt Street
Hingham, MA 02043
Tel. 877-266-5796 (toll free)
Fax 781-740-1677
info@infinitysciencepress.com
www.infinitysciencepress.com

This book is printed on acid-free paper.

O. Beucher and M. Weeks. *Introduction to MATLAB & Simulink: A Project Approach, Third Edition.*
ISBN: 978-1-934015-04-9

The publisher recognizes and respects all marks used by companies, manufacturers, and developers as a means to
distinguish their products. All brand names and product names mentioned in this book are trademarks or service marks
of their respective companies. Any omission or misuse (of any kind) of service marks or trademarks, etc. is not an
attempt to infringe on the property of others.

Library of Congress Cataloging-in-Publication Data

Beucher, Ottmar.   Introduction to MATLAB & SIMULINK : a project approach / Ottmar Beucher and Michael Weeks.
— 3rd ed.
    p. cm.
  Includes bibliographical references and index.
  ISBN 978-1-934015-04-9 (hardcover with cd-rom : alk. paper)
  1. Engineering mathematics–Data processing. 2. Computer simulation–Computer programs.
  3. MATLAB. 4. SIMULINK. I. Weeks, Michael. II. Title.
  TA345.B4822 2007
  620.001'51–dc22
                              2007010556
Printed in the United States of America
7 8 9 4 3 2 1

Our titles are available for adoption, license or bulk purchase by institutions, corporations, etc. For additional
information, please contact the Customer Service Dept. at 877-266-5796 (toll free).

Requests for replacement of a defective CD-ROM must be accompanied by the original disc, your mailing address,
telephone number, date of purchase and purchase price. Please state the nature of the problem, and send the information
to INFINITY SCIENCE PRESS, 11 Leavitt Street, Hingham, MA 02043.

The sole obligation of INFINITY SCIENCE PRESS to the purchaser is to replace the disc, based on defective materials or
faulty workmanship, but not based on the operation or functionality of the product.

CONTENTS

LIST OF FIGURES

LIST OF TABLES

PREFACE

his book is primarily intended for first semester engineering students who are looking for an introduction to the MATLAB and Simulink environment oriented toward the knowledge and requirements of beginning students. Thus, only a few basic ideas from mathematics, in particular ordinary differential equations, programming, and physics are required to understand the contents of this book. This knowledge is usually acquired in the first two or three semesters of a technical engineering degree program.

Under these conditions, this book should also be of interest for practicing engineers who are looking for a brief introduction to MATLAB and Simulink. In the case of this book, engineers will have the knowledge required to understand it years after they have finished their studies.

The MathWorks periodically updates MATLAB and Simulink software. See Appendix C for information about R2007b, the release made in September, 2007. The examples in this book are compatible with the new version.

THE LAYOUT OF THIS BOOK

The first chapter covers the basic principles of MATLAB. It explains the fundamental concepts, how to handle the most important commands and operations, and the basics of MATLAB as a programming language. This chapter, like Chapter 2, emphasizes the numerical solution of ordinary differential equations. Beyond this, there are some comments on the *symbolics toolbox*, which makes the core of the computer algebra program MAPLE available to the MATLAB user and, thereby, enables symbolic calculations in MATLAB.

Chapter 2 is an introduction to the use of Simulink. Here the emphasis is on the solution of ordinary differential equations or systems of differential equations and, with that, the simulation of dynamic systems. Particular attention is paid to the various techniques for interaction between MATLAB and Simulink. Thus, for example, it is shown how the execution of Simulink simulations can be automated in MATLAB.

Both Chapters 1 and 2 are supplemented by a large number of problems which are set up so that they can be worked out by readers as they proceed, *before* starting the next section of the book. The problems are an integral component of the sections and should definitely be worked out independently on a computer right away, because this is the only way to master the subject matter.

Chapter 3 adds a set of programming projects. These are modeled from real-world problems, and go into greater depth than the earlier practice problems. Each section presents a task to be accomplished, then walks the reader through any background information and the solution. The MATLAB code for these projects can be found on the CD-ROM.

In Chapter 4 the problems posed in the first two chapters are provided with *complete solutions*. All the solutions, along with the sample programs discussed in these chapters, are included as files in the *accompanying software*, so that the reader's own computer solutions of the problems can always be checked.

The CD-ROM contains source code for the MATLAB problems and projects, as well as Simulink model files. The sound files generated in Chapter 3 are stored on the CD-ROM under the folder **sweep_data**. It also contains images from the book. Hence, the book is also suitable for self study.

COMMENTS ON THE THIRD EDITION

There are many changes in the new working interface of MATLAB and Simulink. And, naturally, the corresponding discussion has had to be modified. The most important content changes are in the sections on *cell arrays* and *function handles*.

To be sure, cell arrays showed up in the earlier MATLAB versions, but these data structures were not mentioned in the two earlier editions. In the meantime, I have been persuaded to refer to this very flexible tool, even though dealing with these data structures might still be difficult for the beginner. In many cases, however, cell arrays cannot be avoided, so at least the essentials should be discussed thoroughly.

Function handles also showed up earlier. But the concept of calling is new. This makes using the function **feval** superfluous and makes the call natural. This is also discussed in a short section.

Other innovations have not been included, since they go beyond the scope of a basic introduction.

All the new topics have again been supplemented with corresponding problems. Thus, the number of problems (as ever, fully solved) now exceeds 100.

Chapter 3 is also a new addition, with in-depth programming projects.

In sum, the third edition gives the reader has a fully-up-to-date introduction to the current versions of MATLAB and Simulink.

REMARKS ON NOTATION

In this book, MATLAB code is generally set in `typewriter` font. The same holds for MATLAB commands belonging to the built-in MATLAB environment, such as the commands `whos` or the function `ode23`.

MATLAB commands based on the programs written by the authors are also set in `typewriter` font, e.g., the command `FInput`.

Simulink systems belonging to the built-in environment, such as the parameters of Simulink systems, are also generally set in typewriter font, e.g., the parameter `Amplitude` of the Simulink block `Sine Wave`.

The names of Simulink systems provided by the author always begin with an **s_**. This is of historical origin. Before MATLAB 5, MATLAB programs and Simulink systems were m files. Since Simulink systems after MATLAB 5 have the ending °.mdl, it was basically no longer necessary to distinguish them by putting an **s_** in front. But, a second indication of the difference cannot hurt, so this naming convention has been retained.

Metanames, i.e., names in commands, which are to be replaced by the actual name in calls, are set in `<...>`. The formulation `help <command name>` thus means that on calling, the entry `<command name>` must be replaced by the actual command about which help is being sought. Here the angle brackets do not have to be entered.

The names of the accompanying programs are accentuated in `typewriter font` in the text. There is, of course, much more extensive comment in the programs than in the printed excerpts. This is particularly so for the solutions to the problems which have been shortened for reasons of space.

In order to make it easier for the reader to find the programs in the text, an index of the accompanying software is given at the end of the book.

ACKNOWLEDGMENTS

First of all, thanks to my colleagues Helmut Scherf and Josef Hoffmann for their many comments and discussions on this topic.

I also thank the students in the Automotive Technology major in the Department of Mechatronics, who served as "guinea pigs" in some of the one-week compact courses that preceded the development of this book. Naturally, the (known and unknown) reactions of the students had a great influence on the development of this book. I thank Dietmar Moritz, as a representative of them all, for some valuable comments which have been directly incorporated into the book.

O.B., Lingenfeld and Karlsruhe, Germany, 2007

Thanks to the American Institute of Physics, who performed the translation.

I appreciate the support of my students at Georgia State University, especially those who used MATLAB in my Digitial Signal Processing and Introduction to MATLAB Programming classes.

Thanks to Laurel Haislip for playing the violin for the example sound file. I regret that I could only share 17 seconds of your music with the world.

Finally, I would like to acknowledge the support of my wife, Sophie.

M.C.W., Atlanta, Georgia, USA, 2007

Chapter 1

INTRODUCTION TO MATLAB

T his chapter presents the fundamental properties and capabilities of the computational and simulation tool MATLAB.

The purpose of this chapter is to introduce beginners to MATLAB and familiarize them with the basis structure of this software. To make this introduction more accessible, only a few elementary mathematical concepts from linear algebra (vector and matrix calculations) and the analysis of elementary functions will be assumed.

At this point we will avoid more advanced concepts, especially those provided by the MATLAB function libraries (the so-called toolboxes), since they require more extensive knowledge of mathematics, signal processing, control technology, and many other disciplines, and are therefore inappropriate for the beginning students for whom this introduction is intended.

1.1   WHAT IS MATLAB?

MATLAB is a *numerical computation and simulation tool* that was developed into a commercial tool with a user friendly interface from the numerical function libraries LINPACK and EISPACK, which were originally written in the FORTRAN programming language.

As opposed to the well-known computer algebra programs, such as MAPLE or MATHEMATICA, which are capable of performing *symbolic* operations and, therefore, calculating with mathematical equations as a person would normally do with paper and pencil, in principle MATLAB does *purely numerical* calculations. Nevertheless, computer algebra functionality can be achieved within the MATLAB environment using the so-called "symbolics" toolbox. This capability is a permanent component of MATLAB 7 and is also provided in the student version of MATLAB 7. It involves an adaptation of MAPLE to the MATLAB language. We shall examine this functionality in Section 1.8.

Computer algebra programs require complex data structures that involve complicated syntax for the ordinary user and complex programs for the programmer. MATLAB, on the other hand, essentially only involves *a single data structure*, upon which all its operations are based. This is the *numerical field*, or, in other words, *the matrix*. This is reflected in the name: MATLAB is an abbreviation for MATrix LABoratory.

As MATLAB developed, this principle gradually led to a universal programming language. In MATLAB 7 far more complex data structures can be defined, such as the data structure **structure**, which is similar to the data structure **struct** of the C++ programming language, or from the so-called *cell arrays* to the definition of classes in object oriented programming.[1]

Except for structures and cell arrays, which we discuss in Sections 1.3.1 and 1.3.2, in this elementary introduction we will not consider the more advanced capabilities of MATLAB programming, such as object oriented programming and defining certain classes. This would require an extensive background in programming beyond the scope of the present introduction.

If one limits oneself to the basic data structure of the matrix, then MATLAB syntax remains very simple and MATLAB programs can be written far more easily than programs in other high level languages or computer algebra programs. A command interface created for interactive management without much ado, plus a simple integration of particular functions, programs, and libraries supports the operation of this software tool. This also makes it possible to learn MATLAB rapidly.

As we have already noted, MATLAB is not just a numerical tool for evaluation of formulas, but is also an independent programming language capable of treating complex problems and is equipped with all the essential constructs of a higher programming language. Since the MATLAB command interface involves a so-called *interpreter* and MATLAB is an *interpreter language*, all commands can be carried out directly. This makes the testing of particular programs much easier.

Beyond this, MATLAB 7 is also equipped with a very well conceived editor with debugging functionality (Section 1.7), which makes the writing and error analysis of large MATLAB programs even easier.

The last major advantage is the interaction with the special toolbox *Simulink*, which we shall introduce in Chapter 2. This is a tool for constructing simulation programs based on a graphical interface in a way similar to block

---

[1]All data structures (there are 15 different kinds of them) can usually be subsumed under the concept of a "field" (array). Thus, MATLAB yields ARRLAB (ARRay LABoratory). From this concept it follows that the numerical field and, therefore, the classical matrix, is essentially just a special case.

diagrams. The simulation runs under MATLAB and an easy interconnection between MATLAB and Simulink is ensured. In Chapter 2 we shall discuss these and other properties of Simulink in detail.

1.2 ELEMENTARY MATLAB CONSTRUCTS

Starting with the basic data structure of the numerical field, the most important elementary constructs and operations of MATLAB (in the author's opinion) will be presented here. Initially, MATLAB will only be used interactively. It will be shown how (numerical) calculations are carried out interactively and how the results of these calculations can be represented graphically and checked.

The *elementary* MATLAB operations can be divided roughly into five classes:

- Arithmetic operations
- Logical operations
- Mathematical functions
- Graphical functions
- I/O operations (data transfer)

In essence, all these operations involve *operations on matrices and vectors*. These in turn are kept as *variables* which, with very few restrictions, can be defined freely in the MATLAB command interface.

The MATLAB command interface shows up at the start of MATLAB in the form shown in Fig. 1.1 or something similar.[2]

The main elements of this command interface are:

1. the command window,
2. the command-history window,
3. the current directory window or (hidden in this view) the workspace (variable window),
4. the file information window,
5. the icon toolbar with the choice menu for the current directory,
6. the shortcut toolbar, and
7. the start button.

---

[2]This depends on the user's settings. These settings can be specified in the menu command `File - Preferences`.

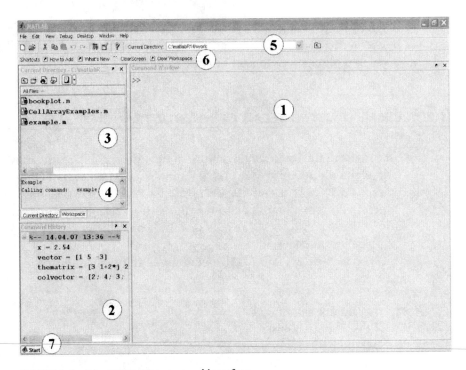

FIGURE 1.1   **The MATLAB command interface.**

The function of the individual elements will be discussed later, in the relevant sections. At present, only the command window, which displays the most important user interface during interactive operation, is of significance.

In the command window (1) MATLAB awaits the user's commands, which are to be directly interpreted and executed by MATLAB following an *input prompt* **>>** (in the student version **EDU>>**). Thus, the user normally communicates with the MATLAB system *interactively*. We shall deal with the possibility of programming in MATLAB in Section 1.6.

Before proceeding to the details of the individual operational classes, the concept of MATLAB variables should be explained further. In doing so, we shall also point out some peculiarities of MATLAB syntax and the interaction with the MATLAB command interface.

1.2.1 MATLAB Variables

A MATLAB variable is an object belonging to a specific data type. As noted at the beginning, the most basic data type is one from which MATLAB takes its name, the *matrix*. For the sake of simplicity, from here on a MATLAB variable is basically *a matrix*. Matrices can be made up of real or complex numbers, as well as characters (ASCII symbols). The latter case is of interest in connection with the processing of *strings* (text). But for now we'll postpone that discussion.

Defining MATLAB Variables

In general, the matrix is defined in the MATLAB command interface by input from the keyboard and assigned a freely chosen variable name in accordance with the following syntax:

```
>> x = 2.45
```

With this instruction, after a MATLAB prompt the number 2.45 (a number is a 1×1 matrix!) will be assigned to the variable **x** and can subsequently be addressed under this variable name.

MATLAB responds to this definition with

```
x =

    2.4500
```

and, in the interactive mode, confirms the input. An error message appears if there are any errors of syntax.

Numbers are always represented by 4 digits after the decimal point by default (format **short**). The default can be changed in the menu command **File - Preferences ...** under the listing **Command - Window - Numeric Format**. In most cases, however, the default representation is the best choice.

The following commands define a row vector of length 3 and a 2×3 matrix. The response of MATLAB is also shown for each case:

```
>> vector = [1 5 -3]

vector =

    1     5    -3

>> thematrix = [3 1+2*i 2; 4 0 -5]

thematrix =

    3.0000        1.0000 + 2.0000i    2.0000
    4.0000                       0   -5.0000
```

Note that the matrix `thematrix` contains a *complex number* as an element. Complex numbers can be defined this way in the algebraic representation using the symbols `i` and `j` that have been reserved for this purpose. Therefore, if possible, these symbols should not be used for other variables.

As the above example shows, the delimiters for the entries in the columns of the matrix are spaces (or, alternatively, commas) and the rows are delimited by semicolons. A column vector can, therefore, be defined as follows:

```
>> colvector = [2; 4; 3; -1; 1-4*j]

colvector =

    2.0000
    4.0000
    3.0000
   -1.0000
    1.0000 - 4.0000i
```

If no variable name is set, MATLAB assigns the name `ans` (answer) to the result, as the following example shows:

```
>> [2,3,4; 3,-1,0]

ans =

    2    3    4
    3   -1    0
```

The Workspace

All of the defined variables will be stored in the so-called *workspace* of MATLAB. You can check the state of the workspace at any time. The command `who` returns the names of the saved variables and the command `whos` provides additional, perhaps vital, information, such as the *dimension of a matrix* or the memory allocation.

This command yields the following for the preceding examples:

```
>> who

Your variables are:

ans        colvector  thematrix  vector     x
```

```
>> whos
  Name            Size        Bytes  Class

  ans             2x3            48  double array
  colvector       5x1            80  double array (complex)
  thematrix       2x3            96  double array (complex)
  vector          1x3            24  double array
  x               1x1             8  double array

Grand total is 21 elements using 256 bytes
```

A very practical way of getting an overview of the content of the workspace is provided by the *workspace browser*, which can be selected using the menu command **Desktop - Workspace**. In Fig. 1.1 the workspace browser has already been opened and is firmly docked in the command interface. In this configuration the workspace is hidden by the window of the current listing (3, 4), but it can be brought into the foreground by clicking the corresponding icon. With a click on the arrow in the menu, you can "undock"[3] the window from its dock in the command interface. This *docking mechanism* can also be applied to all the windows. The extent to which this is used depends on the user's taste.

Fig. 1.2 shows how the workspace browser displays the instructions specified above in its undocked state.

A double click on a variable opens the *array editor* and displays the contents of the variables in the style of a Microsoft Excel Table (Fig. 1.3). Multiple variables can be selected simultaneously (by holding down the control key and clicking) and then the toolbar can be opened with the *open selection* button.[4]

The dimensions of the matrices, the display format, and the individual entries for the matrices can be changed in the array editor. This is especially useful for large matrices, which cannot be viewed in the command window (as in Problem 6). Entire columns or rows can even be copied, erased, or added. In this way it is very easy, for example, to exchange data between Excel and the array editor (and, thereby, to MATLAB) manually with the aid of

---

[3]This action toggles the docking function, and can be used to dock the window there again.

[4]The significance of the button is most easily understood by bringing the mouse pointer over the button and holding it there for a moment. This opens up a text window with the name of the button.

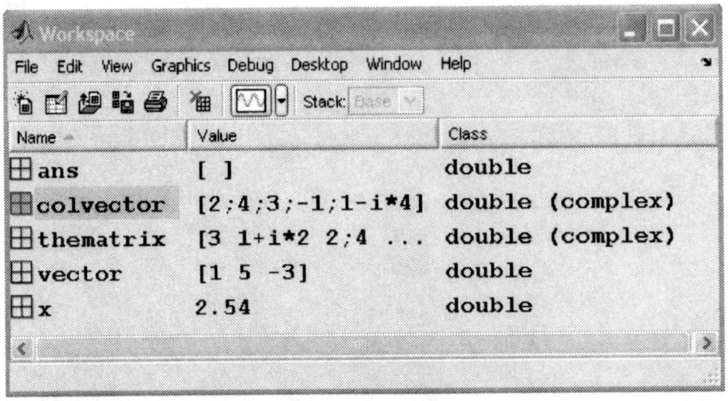

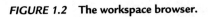

FIGURE 1.2   **The workspace browser.**

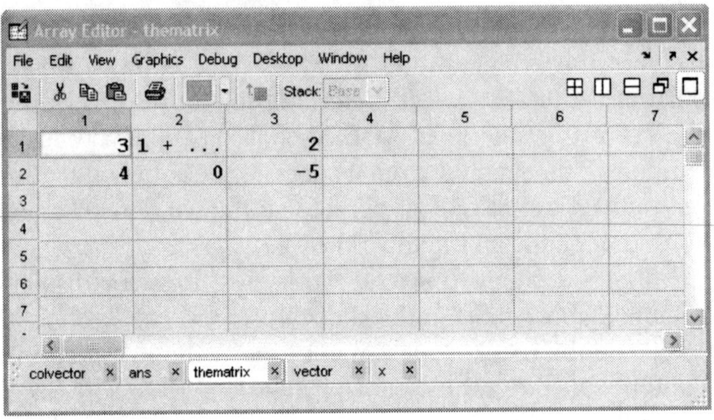

FIGURE 1.3   **Representation of a matrix in the array editor.**

the standard Windows `copy-paste` mechanism in the Windows platform (Problem 6).

If some variables are no longer needed, it is easiest to erase them with the command *clear* in the command interface. For example,

```
>> clear thematrix
```

yields

```
>> who

Your variables are:

ans        colvector  vector    x
```

or the erasure of `thematrix`. The entire workspace is erased using `clear` or `clear all`. These operations are also possible in the workspace browser.

Reconstructing Commands

The commands that have been set previously are saved. In this way you can conveniently *repeat* or *modify a command*. For this only the arrow keys ↑ and ↓ have to be pressed. The earlier commands show up in the MATLAB command window and can (if necessary, after modification) be brought up again using the return key. For instance, the definition of the recently erased matrix `thematrix` can be recovered and reconstructed in this way.

In long MATLAB sessions this keying through earlier commands becomes rather tedious. If you know the first letters of the command, you can shorten the search. Thus, for example, to find the matrix `thematrix` you need only type

```
>> them
```

and then press the arrow key ↑. Only commands beginning with *them* will be searched for. If the beginning is unique, the command is found at once.

Other convenient possibilities employing the so-called *history-mechanism* are provided by the *command-history* window.

In Fig. 1.1 this command-history window (2) can be seen at the lower left. In this window the commands set in the past are listed under the corresponding date of the MATLAB session. By scrolling, it is also very easy to find commands from even further back. A double click on the command is sufficient to activate it again. Some other application possibilities will be discussed in Section 1.4.

The choice of command reconstruction capabilities ultimately depends on the preference of the user and practical considerations.

Other Possibilities for Defining Variables

The problem often arises of expanding a matrix or vector by adding extra components or of eliminating columns and rows.

An expansion can be done in the way described above by adding to the variable name. Thus, the matrix `thematrix` can be expanded by an extra row using the following command:

```
>> thematrix = [thematrix; 1 2 3]
```

```
thematrix =

    3.0000        1.0000 + 2.0000i   2.0000
    4.0000                0          -5.0000
    1.0000        2.0000             3.0000
```

Another column could be added using the following command:

```
>> thematrix = [thematrix, [1;2;3]]

thematrix =

    3.0000        1.0000 + 2.0000i   2.0000   1.0000
    4.0000                0          -5.0000  2.0000
    1.0000        2.0000             3.0000   3.0000
```

or with

```
>> v = [1;2;3]

v =

    1
    2
    3

>> thematrix = [thematrix, v]

thematrix =

    3.0000        1.0000 + 2.0000i   2.0000   1.0000
    4.0000                0          -5.0000  2.0000
    1.0000        2.0000             3.0000   3.0000
```

If you want to delete the second column, then you must fill it with an *empty vector* [ ]. The second column in the variable **thematrix** will then be addressed in accordance with the conventional matrix indexing. Since the row index is arbitrary in this case, it is indicated by the place holder **:** (colon). This yields

```
>> thematrix(:,2) = []

thematrix =

    3     2     1
    4    -5     2
```

```
    1    3    3
```

The first row can thus be deleted using

```
>> thematrix(1,:) =[]

thematrix =

    4    -5    2
    1     3    3
```

Likewise, a row or column vector can be selected and another variable can be assigned. Thus, with

```
>> firstrow = thematrix(1,:)

firstrow =

    4    -5    2
```

the first row of the remaining residual matrix is selected.

Rather than carrying out specified commands in the command window, the operations described above can, of course, also be performed within the array editor. However, with a little practice, working in the command plane is significantly faster. In addition, these operations can also be applied within MATLAB programs; that is, in a *noninteractive* mode (Section 1.6). They are then the only way of obtaining the desired results. Thus, these techniques are of great importance for using the functionality of MATLAB as a programming language.

When it is necessary to process *large matrices or vectors*, outputting the results is often very inconvenient. An example is the following definition of a vector consisting of the numbers from 1 to 5000 in steps of 2. It can be defined simply in MATLAB by the following command, which specifies the starting value, step size, and final value:

```
>> largevector = (0:2:5000)

largevector =

  Columns 1 through 12

     0    2    4    6    8   10   12   14   16   ...

  Columns 13 through 24

    24   26   28   30   etc.
```

Of course, the vector is not displayed here in full.

In order to suppress the output of a MATLAB calculation, the command must be ended with a semicolon. The statement

```
>> largevector = (0:2:5000);
```

lets MATLAB proceed without response in the above case.

If it is desired not to suppress the MATLAB answer entirely, but to show the result on the screen, the command

```
>> more on
```

will let the output in the full command window be suspended when the user ends the execution of the screen display by pressing a key. This function then enables the further progress of the interactive MATLAB session. But it can also be shut off again by entering

```
>> more off
```

as a command.

PROBLEMS

All the important constructs for defining MATLAB variables have now been brought together in order to begin with the first problems on this subject. NOTE ▶ Solutions to all problems can be found in Chapter 4.

Problem 1

Define the following matrices or vectors according to MATLAB and classify the corresponding variables:

$$M = \begin{pmatrix} 1 & 0 & 0 \\ 0 & j & 1 \\ j & j+1 & -3 \end{pmatrix},$$

$$k = 2.75,$$

$$\vec{v} = \begin{pmatrix} 1 \\ 3 \\ -7 \\ -0,5 \end{pmatrix},$$

$$\vec{w} = \begin{pmatrix} 1 & -5.5 & -1.7 & -1.5 & 3 & -10.7 \end{pmatrix},$$

$$\vec{y} = \begin{pmatrix} 1 & 1.5 & 2 & 2.5 & \cdots & 100.5 \end{pmatrix}.$$

Problem 2

1. Expand the matrix M to a 6×6 matrix V of the form

$$V = \begin{pmatrix} M & M \\ M & M \end{pmatrix}.$$

2. Delete row 2 and column 3 from the matrix V (reduced matrix V23).
3. Create a new vector $z4$ from row 4 of the matrix V.
4. Modify the entry $V(4, 2)$ in the matrix V to $j + 5$.

Problem 3

From the vector

$$\vec{r} = \begin{pmatrix} j & j+1 & j-7 & j+1 & -3 \end{pmatrix}$$

construct a matrix N consisting of 6 columns where each contain $\vec{r}$.

Problem 4

Check whether the row vector of Problem 3 can be joined onto the matrix N that was constructed there.

Problem 5

Erase all variables from the workspace and reconstruct the matrix V of Problem 2 using the saved commands and the ↑ and ↓ keys.

Carry out the same procedure once again with the aid of the command-history window.

Problem 6

Fill row 5 of the matrix V of Problem 2 with zeroes using the array editor.

Problem 7

Open the file `ExcelDatEx.xls` of the accompanying software in Microsoft Excel. Transfer the data contained in it to the array editor using the Windows interface.

Next, delete the second column of the transferred data matrix in the MATLAB command window and then copy it back into Excel via the Windows interface.

1.2.2 Arithmetic Operations

The arithmetic operations $(+, -, *, \text{etc.})$ in MATLAB have an important characteristic to which the beginner must become accustomed early on.

Matrix Operations

Since the fundamental data structure of MATLAB is the *matrix*, these operations must, above all, be understood as *matrix operations*. This means that the *computational rules of matrix algebra* are assumed, with all the associated consequences.

Thus, for example, the product of two variables **A** and **B** is *not defined* according to MATLAB if the underlying matrix product $A \cdot B$ is not defined; that is, if the number of columns in **A** is not equal to the number of rows in **B**.

An exception to this rule occurs only if one of the variables is a 1×1 matrix, or *scalar*. Then the multiplication is interpreted as *multiplication by a scalar* in accordance with the rules of linear algebra.

The following examples of MATLAB commands[5] will make this clearer:

```
>>  M = [1 2 3; 4 -1 2]            % define 2x3-Matrix M

M =

      1     2     3
      4    -1     2

>> N = [1 2 -1 ; 4 -1 1; 2 0 1]   % define 3x3-Matrix N

N =

      1     2    -1
      4    -1     1
      2     0     1

>> V = M*N                        % trying the product M*N

V =

     15     0     4
      4     9    -3

>> W = N*M                        % trying the product N*M
??? Error using ==> mtimes
Inner matrix dimensions must agree.
```

---

[5]Comments can be introduced after commands using the % symbol. Later on, this will be very useful in writing MATLAB programs. The characters after % in a line will be ignored by MATLAB.

Thus, MATLAB responds to the attempt to multiply the 2×3 matrix M by the 3×3 matrix N with the 2×3 product matrix V. The attempt to switch the factors fails, since the product of a 3×3 matrix and a 2×3 matrix is *not defined*. MATLAB quits this with the error message `inner matrix dimensions must agree`, which even experienced MATLAB programmers will encounter again and again.

Field Operations

Besides the matrix operations, in many cases there is a need for corresponding arithmetic operations, which must be carried out *term-by-term* (componentwise).

Operations that are to be understood as *term-by-term*, which are known as *field operations* or *array operations* in MATLAB, must at the very least be given a new notation as they might otherwise be confused with matrix operations. This is solved in MATLAB syntax by placing a period (`.`) before the operator symbol. An `*` alone, therefore, always denotes matrix multiplication, while a `.*` always denotes term-by-term multiplication (array multiplication). This leads to other rules about the dimensionality of the objects, as the following example shows:

```
>> M = [1 2 3; 4 -1 2]          % define 2x3-Matrix M

M =

      1      2      3
      4     -1      2

>> N = [1 -1 0; 2 1 -1]          % define 2x3-Matrix N

N =

      1     -1      0
      2      1     -1

>> M*N                           % Matrix product M*N
??? Error using ==> mtimes
Inner matrix dimensions must agree.

>> M.*N                          % term-by-term multiplication
```

```
ans =

     1    -2     0
     8    -1    -2
```

As you can see, the matrix product M*N is not defined this time. Instead, the product is defined *term-by-term*. To each entry in M there is a corresponding entry in N, with which the product can be formed.

Another common example is the squaring of the components of a vector:

```
>> vect = [ 1, -2, 3, -2, 0, 4]

vect =

     1    -2     3    -2     0     4

>> vect^2                        % squaring the vector vect
??? Error using ==> mpower
Matrix must be square.

>> vect.^2                       % vect-squared, term-by-term

ans =

     1     4     9     4     0    16
```

In the first case MATLAB could again carry out a matrix operation. Squaring a matrix, however, is only possible if the matrix is square (i.e., it has the same numbers of columns and rows) which is not so here. But, the *components* themselves can be *squared* in any case.

Division Operations

The division sign has special significance. In MATLAB we distinguish between *right division* / and *left division* \ .

The fact that there are two division operations is again a consequence of the matrix algebra interpretation. *Division of two matrices* A *and* B (i.e., A/B) is *normally not defined*. In the case of square matrices, the quotient can only be interpreted meaningfully as $X = A \cdot B^{-1}$ if the inverse matrix B^{-1} exists. If A^{-1} exists, then "left division" $X = A \backslash B$ is also meaningful, interpreted in this case as $X = A^{-1} \cdot B$.

MATLAB takes these two situations into account by defining left and right division. Let us clarify this with a simple example involving two *invertable* 2×2 matrices.

```
>> A = [2 1 ;1 1]           % Matrix A

A =

     2     1
     1     1

>> B = [- 1 1;1 1]          % Matrix B

B =

    -1     1
     1     1

>> Ainv = [1 -1; -1, 2]     % inverse of A

Ainv =

     1    -1
    -1     2

>> Binv = [-1/2 1/2; 1/2, 1/2]   % inverse of B

Binv =

   -0.5000    0.5000
    0.5000    0.5000

>> X1 = A/B                 % right division

X1 =

   -0.5000    1.5000
         0    1.0000

>> X2 = A*Binv              % checking

X2 =

   -0.5000    1.5000
         0    1.0000
```

```
>> Y1 = A\B                    % left division

Y1 =

    -2.0000   -0.0000
     3.0000    1.0000

>> Y2 = Ainv*B                 % checking

Y2 =

    -2      0
     3      1
```

The next example shows, however, that MATLAB goes one step further in the interpretation of the division operations:

```
>> A = [2 1 ;1 1]             % Matrix A

A =

     2     1
     1     1

>> b=[2; 1]                    % column vector b

b =

     2
     1

>> x = A\b                     % left division of A "by" b

x =

    1.0000
    0.0000

>> y = A/b                     % right division of A "by" b
??? Error using ==> mrdivide
Matrix dimensions must agree.
```

Here "left division," $\vec{x} = A\backslash\vec{b}$, obviously yields a *solution* (in this case unique) *of the linear system of equations* $A\vec{x} = \vec{b}$, as the following test shows:

```
>> A*x

ans =

    2.0000
    1.0000
```

The right division is not defined.

Left and right division can also be used with nonsquare matrices for solving under- and over-determined systems of equations. But since these applications require more extensive mathematical knowledge than can or should be assumed here, we merely note the existence of this possibility.[6]

SUMMARY

Table A.1 of Appendix 1 lists all the arithmetic operations and their execution as *matrix operations*, each with an example.

Table A.2 again lists the arithmetic operations and their execution as *field operations*.

 The MATLAB demonstration program `aritdemo.m` in the accompanying software illustrates the different possibilities associated with arithmetic operations. In particular, it provides practical experience with the difference between matrix and field operations and with the matrix concept in MATLAB. It can be started[7] by copying `aritdemo` to the command window.

PROBLEMS

Work through the following problems for practice with arithmetic operations.
NOTE▶ Solutions to all problems can be found in Chapter 4.

Problem 8
Start the MATLAB demonstration program `aritdemo.m` by calling the command `aritdemo` in the MATLAB command window and work through the program.

---

[6]See the MATLAB 7 handbook.

Problem 9
Calculate

1. the standard scalar product of the vectors

$$\vec{x} = \begin{pmatrix} 1 & 2 & \frac{1}{2} & -3 & -1 \end{pmatrix} \quad \text{and} \quad \vec{y} = \begin{pmatrix} 2 & 0 & -3 & \frac{1}{3} & 2 \end{pmatrix}$$

   using a matrix operation and a field operation,
2. the product of the matrices

$$A = \begin{pmatrix} -1 & 3.5 & 2 \\ 0 & 1 & -1.3 \\ 1.1 & 2 & 1.9 \end{pmatrix} \quad \text{and} \quad B = \begin{pmatrix} 1 & 0 & -1 \\ -1.5 & 1.5 & -3 \\ 1 & 1 & 1 \end{pmatrix},$$

3. using a suitable field operation calculate the matrix

$$C = \begin{pmatrix} -1 & 0 & 0 \\ 0 & 1 & 0 \\ 0 & 0 & 1.9 \end{pmatrix}$$

   from the matrix

$$A = \begin{pmatrix} -1 & 3.5 & 2 \\ 0 & 1 & -1.3 \\ 1.1 & 2 & 1.9 \end{pmatrix}.$$

Problem 10
Test left division $A\backslash\vec{b}$ with the matrix

$$A = \begin{pmatrix} 2 & 2 \\ 1 & 1 \end{pmatrix}$$

and the vector

$$\vec{b} = \begin{pmatrix} 2 \\ 1 \end{pmatrix}$$

and interpret the result.

Problem 11
Test right division $\vec{b}/A$ with the matrix

$$A = \begin{pmatrix} 2 & 2 \\ 1 & 1 \end{pmatrix}$$

and the vector

$$\vec{b} = \begin{pmatrix} 2 & 1 \end{pmatrix}$$

and interpret the result.

Problem 12

Calculate the inverse matrix of

$$M = \begin{pmatrix} 1 & 1 & 1 \\ 1 & 0 & 1 \\ -1 & 0 & 0 \end{pmatrix}$$

using right (or left) division.

1.2.3 Logical and Relational Operations

We shall not go into logical and relational operations in detail here, but only consider the most basic elements. In principle this involves operators that act as *field* operators (i.e., they operate componentwise on the entries in a vector or matrix) and yield logical (truth) values as the result.

For example, you can check whether the components of two matrices in the same terms have an entry $\neq 0$ (=logically true). The following MATLAB sequence shows how this is done using the *logical operator* **&** (logical AND) and what the result of this operation is.

```
>> A=[1 -3 ;0 0]

A =

     1    -3
     0     0

>> B=[0 5 ;0 1]

B =

     0     5
     0     1

>> res=A&B

res =

     0     1
     0     0
```

The resulting matrix res only contains a single 1 (logically true), where the corresponding components of the two matrices are both $\neq$ 0 (logically true), and is 0 (logically false) everywhere else.

The *relational operators* or comparison operators work in a similar fashion. The following sequence checks which components of matrix **A** are greater than the corresponding components of matrix **B**. The matrices from the preceding example are used.

```
>> comp = A>B

comp =

    1    0
    0    0
```

The comparison matrix comp shows that only the first component of **A** is greater than that of **B**.

It hardly needs to be pointed out that here, as with all field operators, the dimensions of the matrices must be identical.

Other relational operators include $\geq$ and $\leq$ (in MATLAB syntax these are <= and >=) and < , as well as == (equal) and ~= (unequal). Besides the above mentioned logical AND, we have the logical operators logical OR (|), logical negation (~), and exclusive OR xor.

Further information on this topic, as well as on the other operations and functions, is available in MATLAB-help (Section 1.5). For comments relating to this section, you can, for example, search for the keyword operators under the menu command Help - MATLAB help. Alternatively, you can enter help ops in the MATLAB command window. In both cases you obtain a list of MATLAB operators that includes the logical and relational operators, among others.

As an illustration of the capabilities of this operator class we give an example that shows up in many simulations. The problem is to *select* each component from a result vector which exceeds a particular value, say 2, and form a vector out of them. This can be done with the following sequence:

```
>> vect=[-2, 3, 0, 4, 5, 19, 22, 17, 1]

vect =

    -2    3    0    4    5    19    22    17    1
```

```
>> compvect=2*ones(1, 9)

compvect =

     2    2    2    2    2    2    2    2    2

>> comp=vect>compvect

comp =

     0    1    0    1    1    1    1    1    0

>> res=vect(comp)

res =

     3    4    5    19   22   17
```

The comparison vector is set up here using the MATLAB command **ones**, which initially yields a matrix containing ones (corresponding to the given components). The comparison with **compvect** yields the places in which the condition (> 2) is satisfied for the vector **vect**. The command **res=vect(comp)** collects these components with the aid of the vector **comp**. In this way only the indices for which this vector $\neq 0$ are employed.

A technique of this sort is often used when values which exceed or lie below a certain threshold (so-called outliers) have to be eliminated from measurement data.

In principle, vectors and matrices that are connected by relational operators must have the same dimension (field operation!). For this reason the vector **compvect** was constructed in the above example. But if, as in this case, the comparison is with only *one* value, this construction can be avoided. In this example,

```
>> comp=vect>2

comp =

     0    1    0    1    1    1    1    1    0
```

also yields the desired result.

It should be noted that the result of a logical operation is *no longer a numerical field*. The input

```
>> whos
  Name          Size              Bytes  Class

  comp          1x9                   9  logical array
  compvect      1x9                  72  double array
  res           1x6                  48  double array
  vect          1x9                  72  double array

Grand total is 33 elements using 201 bytes
```

shows that **comp** is a logical field (*logical array*) (i.e., a field of logical (truth) values). The selection operation **res=vect(comp)** would lead to an error message with a suitably occupied numerical field.

Logical fields do not only arise as a result of comparison operations. They can also be defined directly in the following way:

```
>> logiField1 = [true, true, false, true, false, true]

logiField1 =

     1     1     0     1     0     1

>> numField = [1, 1, 0, 1, 0, 1]

numField =

     1     1     0     1     0     1

>> logiField2 = logical(numField)

logiField2 =

     1     1     0     1     0     1

>> whos
  Name            Size              Bytes  Class

  logiField1      1x6                   6  logical array
  logiField2      1x6                   6  logical array
  numField        1x6                  48  double array

Grand total is 18 elements using 60 bytes
```

The functions **true** and **false** can be used to define entire fields with the logical value "true" or "false." The call without an argument, as above, produces exactly one logical value. Numerical arrays can also be converted into logical arrays using the function **logical**.

As in the above example, components from numerical fields can be selected very simply by means of indexing with logical arrays. Here is another example of this, in which every other component is selected from a vector:

```
>> vect = [1; i; 2; 2+j; -1; -j; 0; i+1]

vect =

    1.0000
         0 + 1.0000i
    2.0000
    2.0000 + 1.0000i
   -1.0000
         0 - 1.0000i
         0
    1.0000 + 1.0000i

>> selct = logical([0 1 0 1 0 1 0 1])

selct =

     0    1    0    1    0    1    0    1

>> selection = vect(selct)

selection =

         0 + 1.0000i
    2.0000 + 1.0000i
         0 - 1.0000i
    1.0000 + 1.0000i
```

PROBLEMS

Work out the following problems to become familiar with logical and relational operations.

NOTE▶ Solutions to all problems can be found in Chapter 4.

Problem 13

Test to see how the logical operations OR (|), negation (˜), and exclusive OR (**xor**) work on the matrices **A** and **B** of Section 1.2.3. If necessary, consult MATLAB-help.

Interpret the result.

Problem 14

Test to see how the relational operations <=, >=, <, ==, and ˜= operate on the vectors

$$\vec{x} = \begin{pmatrix} 1 & -3 & 3 & 14 & -10 & 12 \end{pmatrix} \quad \text{and} \quad \vec{y} = \begin{pmatrix} 12 & 6 & 0 & -1 & -10 & 2 \end{pmatrix}.$$

If necessary, consult MATLAB-help.

Problem 15

Consider the matrix

$$C = \begin{pmatrix} 1 & 2 & 3 & 4 & 10 \\ -22 & 1 & 11 & -12 & 4 \\ 8 & 1 & 6 & -11 & 5 \\ 18 & 1 & 11 & 6 & 4 \end{pmatrix}.$$

Using relational operators set all the terms of this matrix, which are > 10 and < -10 equal to 0.

Hint: First make the comparison and then use the results of this comparison in order to set the appropriate terms equal to 0 using suitable field operations.

Problem 16

Consider the matrix

$$D = \begin{pmatrix} 7 & 2 & 3 & 10 \\ -2 & -3 & 11 & 4 \\ 8 & 1 & 6 & 5 \\ 18 & 1 & 11 & 4 \end{pmatrix}.$$

Use logical fields to select the diagonal of this matrix and save it as a vector **diag**.

1.2.4 Mathematical Functions

MATLAB has an enormous number of different preprogrammed mathematical functions, especially when the functionality of the *toolboxes* is taken into account.

For the beginner, however, only the so-called *elementary functions* are relevant. These include well-known functions, such as the cosine, sine, exponential function, logarithm, etc., from real analysis.

Given the data structure concept of MATLAB, which essentially relies on *matrices*, it might seem as though there might be a problem because these functions are not defined for matrices at all.

Another glance, however, shows that the solution to this problem is immediately obvious in the examples that have already been discussed. Naturally, the action of an elementary function on a vector or a matrix is again only meaningful in a *term-by-term* sense. The following sequence shows what is meant by the sine of a vector.

```
>> t=(0:1:5)

t =

     0     1     2     3     4     5

>> s=sin(t)

s =

     0    0.8415    0.9093    0.1411   -0.7568   -0.9589
```

The sine of the specified vector $\vec{t}$ is again a vector, specifically, the vector

$$\vec{s} = (\sin(0), \sin(1), \sin(2), \cdots, \sin(5)).$$

The full significance of this technique will only become clear to the MATLAB user in the course of extensive simulations. It should be kept in mind that the above sequence replaces a *programming loop*. Thus, in the C++ programming language, the sequence would be implemented as follows:

```
double s[6];
for(i=0; i<6; i++)
    s[i] = sin(i);
```

The advantage is not obvious in this little example, but the same technique can produce very elegant solutions for large simulations in MATLAB.

At this point we cannot discuss all the elementary functions. An overview of the available functions can be obtained by entering the command **help elfun** in the MATLAB command interface. This gives a list of the names of the functions together with brief descriptions of each.

More direct information is obtained with `help <functionname>`. Thus, `help <asin>` yields

```
>> help asin

 ASIN    Inverse sine.
    ASIN(X) is the arcsine of the elements of X. Complex
    results are obtained if ABS(x) > 1.0 for some element.

    See also sin, asind.

    Overloaded functions or methods
    (ones with the same name in other directories)
       help sym/asin.m

    Reference page in Help browser
       doc asin
```

a description of the arcsine.

Here, we should point out that function names and call syntax in *help* are always in capital letters. This merely serves as emphasis in the help text. When the functions are used in the command window, the commands must generally be written in lowercase letters. Unlike in earlier versions, MAT-LAB 7 distinguishes strictly between upper and lowercase letters (and is *case sensitive*). In the above example the two calls

```
>> x = 0.5;
>> ASIN(X)
??? Undefined function or variable 'X'.

>> X = 0.5;
>> ASIN(X)
??? Undefined command/function 'ASIN'.
```

each yield error messages, since in the first the variable **X** does not exist and in the second the function **ASIN** is unknown.

A correct call looks like

```
>> x = 0.5;
>> asin(x)

ans =

    0.5236
```

Besides the functions from the `elfun` group, the so-called *special mathematical functions*, which are listed using `help specfun`, are also of interest. These, however, require a more profound mathematical knowledge and will not be discussed further here. Problem 22 is recommended for the interested reader.

Two more examples illustrate the manipulation of elementary functions in MATLAB. First, the magnitude and phase, in *radians* and in *degrees*, of a vector of complex numbers is to be determined. This problem is frequently encountered in connection with the determination of the so-called transfer function of linear systems.

```
>> cnum=[1+j, j, 2*j, 3+j, 2-2*j, -j]

cnum =

  Columns 1 through 3

    1.0000 + 1.0000i       0 + 1.0000i        0 + 2.0000i

  Columns 4 through 6

    3.0000 + 1.0000i    2.0000 - 2.0000i       0 - 1.0000i

>> magn=abs(cnum)

magn =

    1.4142    1.0000    2.0000    3.1623    2.8284    1.0000

>> phase=angle(cnum)

phase =

    0.7854    1.5708    1.5708    0.3218   -0.7854   -1.5708

>> deg=angle(cnum)*360/(2*pi)

deg =

   45.0000   90.0000   90.0000   18.4349  -45.0000  -90.0000
```

In the following example all the terms in a matrix (say, many series of measurements of a voltage signal) are to be converted into decibels (dB). For a voltage signal U this means that the value must be converted into

$$20 \cdot \log_{10}(U)$$

so that:

```
>> meas=[25.5 16.3 18.0; ...
         2.0   6.9  3.0; ...
         0.05  4.9  1.1]

meas =

   25.5000   16.3000   18.0000
    2.0000    6.9000    3.0000
    0.0500    4.9000    1.1000

>> dBmeas=20*log10(meas)

dBmeas =

   28.1308   24.2438   25.1055
    6.0206   16.7770    9.5424
  -26.0206   13.8039    0.8279
```

Incidentally, the above example shows how a too-long MATLAB command can be broken up without being erased when the return key is pressed. Just type three periods[8] (...) before pressing the return key and the command can be continued in the next line.

In the next example a list of points in the first quadrant of the cartesian $\mathbb{R}^2$ are to be converted to polar coordinates. The list is in the form of a matrix of (x, y) values. As is well known (see the example in Section 1.6.3), the polar coordinates (r, ϕ) in this case are given by

$$r = \sqrt{x^2 + y^2}, \tag{1.1}$$

$$\phi = \arctan\left(\frac{y}{x}\right). \tag{1.2}$$

---

[8]The periods (...) used to continue the command line are, of course, no longer obligatory in this special case, since MATLAB 7 "takes note" that it is dealing with an incomplete matrix definition as long as the closing bracket is not entered. In earlier MATLAB versions these periods always had to be present. As before, the periods (...) have to be inserted in order to break up a command line.

The calculation is carried out by the following MATLAB sequence, using the elementary mathematical functions **sqrt** and **atan**:

```
>> points = [ 1 2; 4 3; 1 1; 4 0; 9 1]

points =

        1     2
        4     3
        1     1
        4     0
        9     1

>> r = sqrt(points(:,1).^2 + points(:,2).^2)

r =

    2.2361
    5.0000
    1.4142
    4.0000
    9.0554

>> phi = atan(points(:,2)./points(:,1))

phi =

    1.1071
    0.6435
    0.7854
         0
    0.1107

>> polarc = [r,phi]

polarc =

    2.2361    1.1071
    5.0000    0.6435
    1.4142    0.7854
    4.0000         0
    9.0554    0.1107
```

Note the use of the **:** operator and the field operations used for calculating **r** and **phi**, which make it possible to perform the calculation on the whole vector in a single stroke.

Finally, it should be noted that the preceding calculation could be shortened using the following instructions:

```
>> points = [ 1 2; 4 3; 1 1; 4 0; 9 1]

points =

        1     2
        4     3
        1     1
        4     0
        9     1

>> polarc = [sqrt(points(:,1).^2 + points(:,2).^2), ...
                  atan(points(:,2)./points(:,1))]

polarc =

    2.2361    1.1071
    5.0000    0.6435
    1.4142    0.7854
    4.0000         0
    9.0554    0.1107
```

PROBLEMS

Work out the following problems for practice with the elementary mathematical functions.

NOTE   Solutions to all problems can be found in Chapter 4.

Problem 17

Calculate the values of the signal (function)

$$s(t) = \sin\left(2\pi\,5t\right)\cos\left(2\pi\,3t\right) + e^{-0.1t}$$

for a time vector of times between 0 and 10 with a step size of 0.1.

Problem 18
Calculate the values of the signal (function)

$$s(t) = \sin\left(2\pi\,5.3t\right)\sin\left(2\pi\,5.3t\right)$$

for the time vector of Problem 17.

Problem 19
Round off the values of the vector

$$s(t) = 20\sin\left(2\pi\,5.3t\right)$$

for the time vector of Problem 17, once up and once down. Find the corresponding MATLAB functions for this. For this, use the MATLAB help mechanisms. If necessary, first consult Section 1.5.

Problem 20
Round the values of the vector

$$s(t) = 20\sin\left(2\pi\,5t\right)$$

to the nearest integer for the time vector of Problem 17 using a suitable MATLAB function. Also, give the first 6 values of $s(t)$ and the corresponding rounded values in a matrix with two rows and interpret the somewhat surprising result.

Problem 21
Calculate the base 2 and base 10 logarithms of the vector

$$\vec{b} = (1024 \quad 1000 \quad 100 \quad 2 \quad 1)$$

using the appropriate elementary MATLAB functions.

Problem 22
Convert the cartesian coordinates from the above example (in variable `points`) into polar coordinates using the special mathematical function `cart2pol`. First, check on the syntax of this command by entering `help cart2pol` in the command window.

1.2.5 Graphical Functions

One of the outstanding strengths of MATLAB is its capability of *graphical visualization* of computational results.

For this, MATLAB has available very simple but powerful graphical functions. MATLAB is thus able to represent ordinary functions in *two-dimensional plots* (xy plots) and functions of two variables in *three-dimensional plots* with perspective (xyz plots).

A complete survey of all the graphical functions can be obtained by entering `help graph2d`, `help graph3d`, and `help graphics` in the command window or in the corresponding entries of MATLAB `help`. In this section we shall discuss only the most useful and important of these functions.

Two-dimensional Plots

By far the most important and the most commonly used graphical function is the function `plot`. Entering `help plot` provides the following information (excerpt):

```
>> help plot

PLOT    Linear plot.
   PLOT(X,Y) plots vector Y versus vector X. If X or Y is a
   matrix, then the vector is plotted versus the rows or
   columns of the matrix, whichever line up ...
```

MATLAB help (of which we present only part of the full display for reasons of space) shows the basic capabilities of `plot`. Beyond that, there are many other possibilities and options for `plot`. A discussion of this topic is beyond the scope of this book. The interested reader can look at the book by Marchand.[9]

As the help function shows, in principle, two vectors are required in order to plot a graph. The first vector represents the vector of x values and the second, the corresponding vector of y values. The vectors must, therefore, have the *same length*. If this is not so, MATLAB produces an appropriate error message unless x is a scalar (see `help plot`). This message shows up in practice, even for experienced MATLAB users, but the error is easily corrected.

Many functions can also be plotted simultaneously, either by writing the x, y pairs one after the other in the parameter list or, when the same x vector will always be used, by combining the y vectors into a corresponding matrix.

In addition, the form of the graphs in terms of the type of line and color can be varied by using suitable parameters.

Our first example again uses the variables `s` and `t` from the example in Section 1.2.4, and plots the result obtained there:

```
>> t = (0:1:5);
>> s = sin(t);
>> plot(t,s)
```

---

[9]P. Marchand, *Graphics and GUIs with MATLAB*, CRC Press, 1999.

Note that in this example numerical output of the variables **t** and **s** is suppressed.

The plot command opens a window with the plot shown in Fig. 1.4. Here and in the following graphs the scales on the axes are enlarged for the sake of clarity relative to the default. For the same reason, the thickness of the lines is also somewhat greater than the original display. See the corresponding picture in Fig. 1.12.

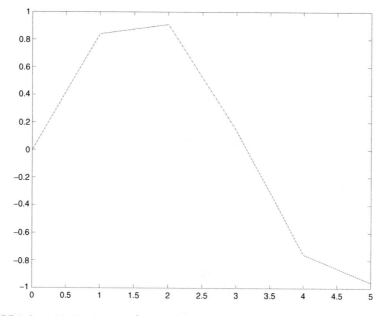

FIGURE 1.4   A MATLAB example of a simple _x, y_ plot.

You can see that the result is normally in the form of a polygonal sequence. This often leads to erroneous interpretations of graphs, especially when the abscissa values are far apart. Be careful. In addition, only numbers are placed on the axes automatically. You have to put in grids, axis labels, and titles yourself using the appropriate MATLAB commands.

The following commands generate the plot in Fig. 1.5. Here another line style (circles) has been chosen and the color has been changed to magenta (which cannot, of course, be shown here). (The default color in MATLAB is blue.)

```
>> t=(0:1:5);
>> s=sin(t);
>> plot(t,s,'k*')
```

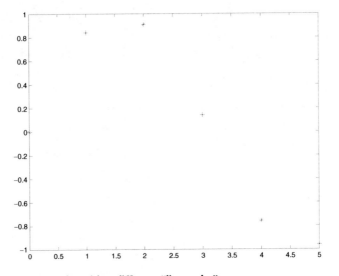

FIGURE 1.5   The *x, y* plot with a different "line style."

In the next example a sine of amplitude 1 with frequency[10] 5 Hz, a cosine of amplitude 2 and frequency 3 Hz, and the exponential function e^{-2t} are calculated for the time vector t=(0:0.01:2) and plotted:

```
>> t=(0:0.01:2);
>> sinfct=sin(2*pi*5*t);
>> cosfct=2*cos(2*pi*3*t);
>> expfct=exp(-2*t);
>> plot(t,[sinfct; cosfct; expfct])
```

This yields the graph shown in Fig. 1.6.

In Fig. 1.7 this graph is shown again, but using some of the possibilities for different line shapes in order to distinguish the plots of the functions from one another. The corresponding plot statement is

```
>> plot(t,sinfct,'k-', t, cosfct, 'b--', t, expfct, 'm.')
```

Here the variable t has to be repeated anew each time. The sine is plotted with the above settings, a black (**black**) solid curve (-), while the cosine is plotted in blue (**blue**) with a dashed curve (- -), and the exponential function is in magenta (**magenta**) with a dotted curve (.).

---

[10]Note that the general form for a harmonic oscillation with frequency f, amplitude A, and phase φ is $f(t) = A \sin(2\pi ft + \varphi)$.

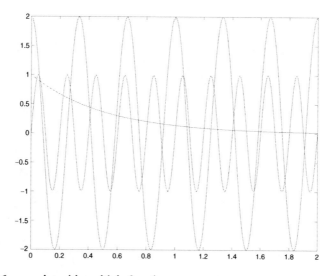

FIGURE 1.6   **x, y plot with multiple functions.**

Besides these most frequently used plot functions, MATLAB also has a number of other two-dimensional plot functions. In signal processing the function **stem** is often used. It presents discrete signals in the form of a fence plot. This function, however, is only suitable for sparse data sets, for otherwise the lines come too close to one another and the small circles at the ends of

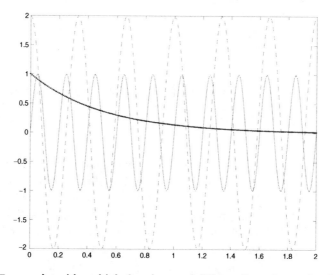

FIGURE 1.7   **x, y plot with multiple functions and different line colors and styles.**

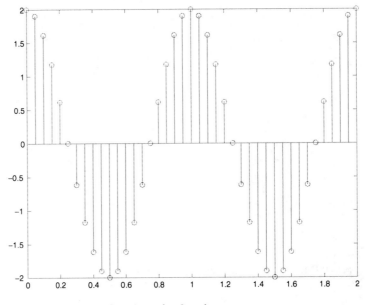

FIGURE 1.8   x, y **plot using the** stem **plot function.**

the lines can overlap in an unsightly manner. In these cases **plot** is more intuitive. Fig. 1.8 shows the results of the following MATLAB sequence:

```
>> t=(0:0.05:2);
>> cosfct=2*cos(2*pi*t);
>> stem(t,cosfct)
>> box
```

One function that is specially suited to displaying time-discrete signals is the function **stairs**, which represents the signal in the form of a stair function. Fig. 1.9 shows the result of the call

```
>> stairs(t,cosfct)
```

for the cosine signal defined above. A suitable labelling of the axes and a title line are also important for the documentation of graphs of this sort. In addition, it is often also necessary to provide a grid for the graph in order better to compare the values or to enlarge segments of a graph.

MATLAB, of course, has functions for this purpose. For the axis labels there are **xlabel**, **ylabel**, and **title**, for the grid there is the function **grid**, and for magnification there is the function zoom, which allows the enlargement of a segment with the mouse, as well as the command **axis**,

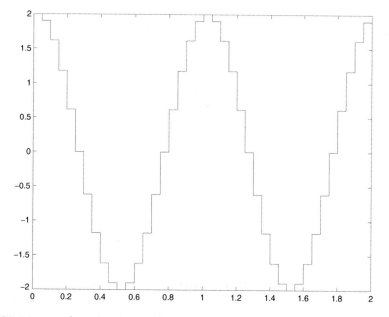

FIGURE 1.9 *x, y* plot using the `stairs` function.

for which the *x* and *y* ranges must be specified explicitly. Labels can be inserted inside the graph using the function **text**.

The graphs in Figs. 1.10 and 1.11 show the result of the following MATLAB command sequence, in which the above capabilities are used:

```
>> t=(0:0.05:2);
>> cosfct=2*cos(2*pi*t);
>> plot(t,cosfct)
>> xlabel('time / s')
>> ylabel('Amplitude / V')
>> text (0.75,0,'\leftarrow zero crossing','FontSize',18)
>> title('A cosine voltage with frequency 1 Hz')
>> figure              % open a new window !
>> plot(t,cosfct)
>> xlabel('time / s')
>> ylabel('Amplitude / V')
>> grid                % set grid frame
>> axis([0, 0.5, 0, 2]) % detail within interval [0, 0.5],
                       % but only amplitudes within
                       % the interval [0, 2]
>> title('Detail of the cosine voltage with frequency 1 Hz')
```

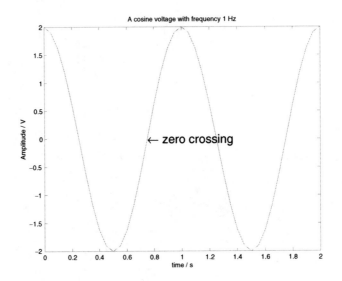

FIGURE 1.10   **A labelled *x, y* plot using the plot function.**

We shall not discuss the other two-dimensional plot functions at this point. The reader can find out more about these possibilities in the help menu or in the book by Marchand[11] and experiment with them.

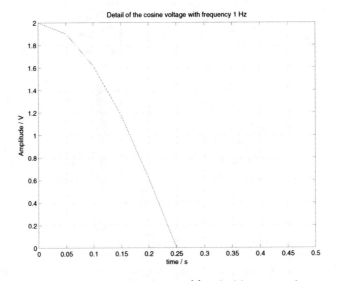

FIGURE 1.11   **A segment of Fig. 1.10 using the `` axis '' command.**

[11]P. Marchand, *Graphics and GUIs with MATLAB*, CRC Press, 1999.

In the following we shall, however, briefly discuss the possibilities for processing graphs, which are specially available during interactive operation. Then three-dimensional graphs will be discussed.

Graphical Processing

MATLAB 7 has extensive improvements in the user interface compared to its predecessors (see Section 1.4). Thus, it offers some convenient possibilities for the processing and export of graphics in the interactive mode. Since this is of great importance for the documentation of MATLAB computations, we shall discuss it briefly at this point.

The plot window of a MATLAB graphic as seen by the user has the form shown in Fig. 1.12.

The essential functions for manipulation of graphics can be selected using the so-called *toolbar* located below the pull-down menus. These functions

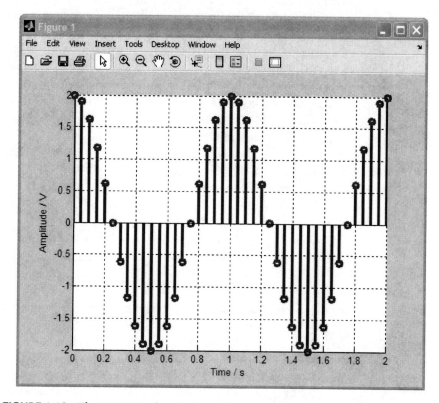

FIGURE 1.12 **The MATLAB plot window with a sample graph.**

can, of course, also be reached through the pull-down menu itself. Besides the conventional Windows icons for save, print, etc., in the left-hand third of the toolbar, in the middle third of the toolbar you can see the magnifier symbol, with which a graphic can be enlarged or shrunk, as well as a symbol for rotating a graphic. The latter is especially practical for viewing three-dimensional graphics (see the next subsection) and is very useful. The icons in the right-hand third of the toolbar provide access to other useful functions, such as the so-called *data cursor*, with which the (x, y) coordinates of a point in the graph can be displayed by clicking on the given point. There is also a button with which the graphics can be provided interactively with a legend. The `show plot tools` icon on the far right extends the plot window to a wider window with tools for graphical processing.

A detailed presentation of the possibilities at this point would go beyond the scope of this introduction. Instead, we shall illustrate the functionality with a small example.

If, for instance, you want to change the title label, select and click on the title text with the plot editing arrow and make your change. Similarly, by double clicking on the lines it is possible to change the line style. A double click on the axes makes it possible to change the axis scale and other properties of the axes. With a mouse click the plot window can be expanded at any time into a configuration window where the desired settings can be chosen. All the configuration possibilities can be obtained in a single view if, as mentioned above, you press the `show plot tools` icon. Fig. 1.13 shows the plot window when it is expanded in this way.

A selection of tools for manipulating a plot are also available. This is otherwise context dependent. Thus, the `property editor - axes` window is opened below when the intersection of the axes is clicked. If the function graph is clicked, then this window will be replaced by the corresponding `property editor` window.

Besides these, many more functions can be obtained through the menu items `insert` and `tools`, which make it possible to modify the appearance of a graphic afterward. This calls on the experimental playfulness of the reader.

Given the innumerable possibilities offered by the plot window for changing graphics in MATLAB 7 in the interactive mode, the preceding discussion of the plot commands may seem to be of somewhat dubious utility for this purpose. In fact, processing plots with the tools in the plot window is simpler and more intuitive. Moreover, no commands have to be learned in combination with their syntax. It has already been pointed out that MATLAB is a programming language (see Section 1.6). If you want to prepare completed

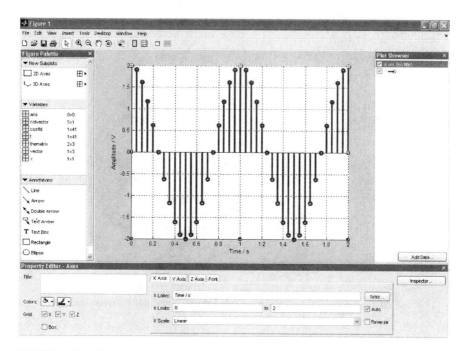

FIGURE 1.13 **The plot-tools window.**

graphics within a program, the above MATLAB commands provide the only way to do it.

Finally, the important functionalities for graphics export from MATLAB 7 should be mentioned at this point. In the plot window various formats, such as *encapsulated postscript*, *TIFF*, *portable network graphics*, etc., into which a graphic can be converted and saved, may be chosen using the menu command `File - Save as ...`. Of course, the graphics can be exported into other applications via the Windows interface using `Edit - Copy figure` command.

Three-dimensional Plots

Here again, we shall not discuss all the capabilities of MATLAB in detail. The most important three-dimensional graphics functions are certainly `mesh` and `surfsurf` for the representation of a three-dimensional mesh or surface-mesh plot and `contour` for a contour plot. In the following example the two-dimensional function

$$f(x,y) = \sin{(x^2 + y^2)}e^{-0.2 \cdot (x^2 + y^2)}$$

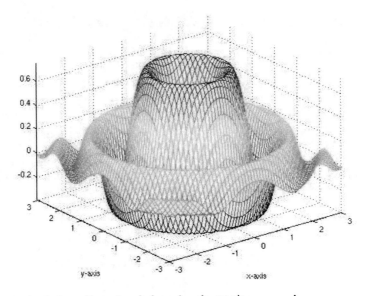

FIGURE 1.14   A three-dimensional plot using the mesh command.

is plotted using the function **mesh** or **surf** within the square

$$[-3,3] \times [-3,3],$$

which is provided with a rectangular grid with a step size (edge length) of 0.1. To do this the value of the function is calculated at every grid point. The value of the function is attached to a surface mesh (tile) and the entire graph is plotted in perspective with respect to a viewpoint specified by MATLAB (which, of course, can easily be changed using the methods described in the preceding section).

The following sequence of commands produces the plots shown in Figs. 1.14 and 1.15:

```
>> x=(-3:0.1:3);          % grid frame in x direction
>> y=(-3:0.1:3)';         % grid frame in y direction
>> v=ones(length(x),1);   % auxiliary vector
>> X=v*x;                 % grid matrix of the x values
>> Y=y*v';                % grid matrix of the y values
>>                        % function value
>> f=sin(X.^2+Y.^2).*exp(-0.2*(X.^2+Y.^2));
>> mesh(x,y,f)            % mesh plot with mesh
>> mxf = max(max(f));     % maximum value of the function
>> mif = min(min(f));     % minimum value of the function
```

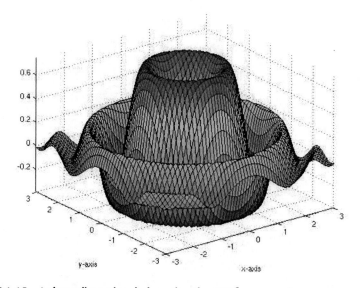

FIGURE 1.15   A three-dimensional plot using the surf command.

```
>> axis([-3,3,-3,3,mif,mxf])% adjust axes
>> xlabel('x-axis');        % label axes
>> ylabel('y-axis');
>> figure                   % new plot
>> surf(x,y,f)              % surface mesh plot with surf
>> axis([-3,3,-3,3,mif,mxf])% adjust axes
>> xlabel('x-axis');        % label axes
>> ylabel('y-axis');
```

The technique for generating the x, y grid is of some interest. Here, elegant use has been made of matrix algebra, avoiding the need to write a double loop as would have to be done in the C++ programming language.

The reader should certainly take Problem 28 to heart in order to grasp what has actually been done and what the trick involves.

In order to compare the plot results from **mesh** and **surf**, the graph of Fig. 1.16 shows a three-dimensional contour plot of the function. It was generated using the following MATLAB commands:

```
>> contour3(x,y,f,30)
>> axis([-3,3,-3,3,mif,mxf]) % adjust axes
>> xlabel('x-axis');         % label axes
>> ylabel('y-axis');
```

The number 30 represents the number of contours to be plotted.

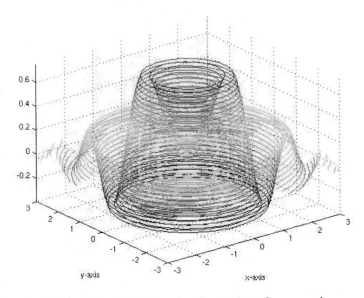

FIGURE 1.16 A three-dimensional plot using the `contour3` command.

Subplots and Overlay Plots

Besides being able to plot graphs in different individual windows, MATLAB can also plot multiple (overlay) graphs in a *single* window. This can be done, for example, using the **subplot** command.

The following example shows how the magnitude and phase of the complex function $f(t) = t^2 e^{jt} j^t$ can be plotted with the two plots one above the other:

```
>> t=(0:0.1:5);
>> f=(t.^2).*exp(j*t).*(j.^t);   % * and ^ are field operations.
>> subplot(211)                  % plot the top graph
>> plot(t,abs(f))
>> subplot(212)                  % plot the bottom graph
>> plot(t,angle(f))
```

Fig. 1.17 shows the result.

Here we see that the command **subplot** only sets the axes of the graph in the right place. The plot, itself, is still provided by the **plot** command.

The parameters of the **subplot** command specify how many graphs are to be drawn altogether in the vertical (the first number) and horizontal (second number) directions. The third number specifies which of the subgraphs

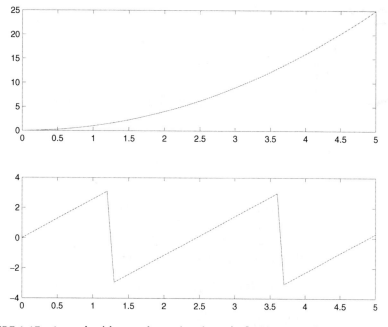

FIGURE 1.17   **A graph with two plots using the** subplot **command.**

is meant, going from upper left to lower right. For example **subplot(325)** signifies the fifth plot in an array of 3×2 graphs (for the next plot command).

Another possibility for multiple plots in a single window is *overlay graphs* as shown in Figs. 1.6 and 1.7. Besides the ways described previously, multiple graphs can be plotted *on top of one another* using the **hold** command. This freezes the current graph so that the next graph is not plotted in its own window, but in the same window. Here is an example:

```
>> t = (0:0.5:10);
>> sinfct = sin(2*pi*5*t);
>> cosfct = 2*cos(2*pi*3*t);
>> plot(t,sinfct)
>> hold
>> plot(t,cosfct)
```

Here, the current graph is always the graph whose plot window is "on top," and, therefore, the last one processed. Before using the **hold** command, you have to make sure that the correct graph is being written over by clicking the corresponding window.

PROBLEMS

Work out the following problems to familiarize yourself with the plot functions.

Solutions to all problems can be found in Chapter 4.

Problem 23
Figure out why the sequence

```
>> t=(0:0.01:2);
>> sinfct=sin(2*pi*5*t);
>> cosfct=2*cos(2*pi*3*t);
>> expfct=exp(-2*t);
>> plot(t,[sinfct, cosfct, expfct])
```

leads to a MATLAB error.

Problem 24
Enter the MATLAB command sequence

```
>> t=(0:0.5:10);
>> sinfct=sin(2*pi*5*t);
>> cosfct=2*cos(2*pi*3*t);
>> expfct=exp(-2*t);
>> plot(t,[sinfct; cosfct; expfct])
```

and interpret the (somewhat odd) graphical result.

Problem 25
Experiment with the plot functions **plot** and **stem** using different discretization step sizes and different line styles.

Vary the line style and axis partition using the editor command in the plot window.

Problem 26
Experiment with the plot functions **semilogx**, **semilogy**, and **loglog**, in order to become acquainted with the different possibilities for displaying logarithmic axes.

For this, use the frequency vector

$$\omega = (0.01,\ 0.02, 0.03,\ 0.04, \dots, 5) \quad \text{rad/s},$$

which shows up in the so-called "transfer functions"[12] of an integrator,

$$H(j\omega) = \frac{1}{j\omega}$$

and of a first order time delay element,

$$H(j\omega) = \frac{1}{1+j\omega} \, ,$$

which are encountered in signal processing and control engineering. Plot the magnitude of these functions.

Decide which display is best.

Problem 27

Label the plot results from Problem 26 using the appropriate MATLAB commands and experiment with the commands **axis** and **zoom**.

For comparison, work out the same problems using the editor commands in the plot window.

In addition, familiarize yourself with the documentation of a MATLAB graphic in a text processing program where you insert (**paste**) the graphic into a document using the interface **Edit - copy figure**.

Problem 28

Test the command sequence for the **surf** plot example in Section 1.2.5 with two vectors of the form

$$\vec{x} = \begin{pmatrix} -1 & 0 & 1 \end{pmatrix} \quad \text{and} \quad \vec{y} = \begin{pmatrix} -2 & 0 & 2 \end{pmatrix}$$

and find the trick used to generate the x, y grid.

Problem 29

Check out some of the other three-dimensional plot functions using the example in Section 1.2.5. You can get a survey of these plot functions by typing in **help graph3d** in the MATLAB command interface.

Problem 30

Plot the magnitude and phase of the transfer functions used in Problem 26 as a overlay plot (to obtain a so-called *Bode diagram*). Use a logarithmic scale for the magnitude on the y axis.

---

[12]The concept of "transfer function" is discussed in the pertinent literature. An understanding of this concept is not essential for the problem under consideration.

Problem 31

Plot the exponential functions

$$e^{-t/2} \quad \text{and} \quad e^{-2t/5}$$

over the interval $[0, 2]$.

1. In a single overlay plot,
2. next to one another in different plots, and
3. one above the other in different plots.

Label one of the plots in a suitable fashion and then try to save this plot in an Microsoft Word file using the interface.

Problem 32

Calculate and plot the function

$$x^2 + y^2$$

on the rectangle $[-2, 2] \times [-1, 1]$.

For this, use a grid that has equidistant step sizes of 0.2 in the x direction and 0.1 in the y direction.

1.2.6 I/O Operations

I/O operations (**Input/Output** operations) relate to data transfer between MATLAB and external files. A mechanism of this sort is necessary if, for example, you want to import externally acquired data into MATLAB or transfer results from MATLAB to other applications.

The `Load` and `Save` Commands

MATLAB recognizes many file interfaces. The most elementary of these are represented by the `load` and `save` commands.

Thus, for example, with the command

```
>> save wsbackup
```

the content of the work space (i.e., that which is indicated by `whos`) will be saved in a binary file named `wsbackup.mat` (ending in `.mat`). For this, the target directory is the work directory set in the menu entry `current directory` (see item (5) in Fig. 1.1). This setting can obviously be overwritten by explicitly specifying a directory in front of the file name.

Besides backup of the workspace, individual variables can also be saved. (See below.)

In MATLAB 7 the binary format has been changed relative to the earlier versions. Under certain conditions this makes the exchange of these files with other MATLAB users difficult. In order to ensure down compatibility, an option (-V6) can be used in the save command in order to save the file in an earlier binary format. The following instruction saves the variables var1 and var2 in MATLAB 6 binary form in a file that will be set up in a (explicitly specified) directory outside the Current directory:

```
>> save 'C:\beucher\matlab7\thevars' var1 var2 -V6
```

To save the workspace variables in MATLAB 6 binary format (which is not recommended in light of the further development of MATLAB), set them in MATLAB desktop menu under File - preferences ... in the General - MAT - files chart.

The variables saved in this way can again be read into the workspace using load <path/filename> or load <filename>.

By specifying additional options, which are described in the help corresponding to the commands, you can also change the file format in which the variables are saved. An important case of this is the ASCII format (printable characters).

Thus, for instance, a variable X in the workspace could be saved in ASCII format as the file thevarX.txt using

```
>> save thevarX.txt X -ASCII
```

Other external ASCII files (e.g., containing series of measurements taken using other programs) could be imported (again) using load. In any case, the data have to be prepared in a *matrix format* in order to do this, so that MATLAB can assign it a matrix (whose name is then automatically identical with the file name without its ending) in the workspace. For example, the variable X saved above using the *save* command can again be loaded into the workspace using

```
>> load -ASCII thevarX.txt
```

and it then shows up there under the name thevarX (but not X) and is again available.

As opposed to save and load, in the *interactive* mode the menu commands File - Save workspace as ... and File - import `data` can be used. In the latter case the so-called *import wizard*, which provides flexible data import possibilities, is opened.

1.2.7 Import Wizard

The `save` and `load` commands are thus, above all, indispensable when loading and saving operations must be carried out *within programs* (see Section 1.6). If MATLAB 7 is used in the interactive mode, that is from the command interface, then the import wizard is available as an alternative for importing data. On choosing the menu command `File - import data` a file selection window is initially opened, in which choices for the working directory and various file formats are provided. When a file is clicked, the dialogue field shown in Fig. 1.18 opens up.

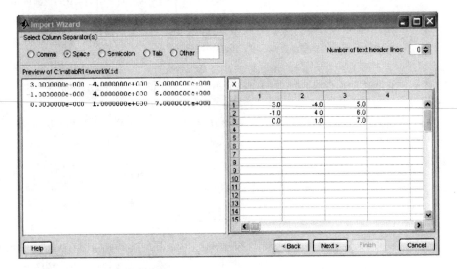

FIGURE 1.18   **The import wizard.**

Here, a preview of the data to be imported is provided and some additional settings may be specified before the matrix is saved in the workspace by clicking the `Finish` button. The save ends with an appropriate message in the command plane.

1.2.8 Special I/O Functions

Besides the above mentioned functions, the I/O operations include a number of functions for writing and reading files, which have a syntax similar to the corresponding C/C++ functions.

Beyond that, functions for importing picture and audio information or for formats of other applications, such as Microsoft Excel, are of special interest.

A complete survey of the functions can be obtained by entering `help iofun`, `help audiovideo`, and `help imagesci`. These commands list a whole series of MATLAB functions with which the import and export of various file formats can be accomplished. A discussion of all of these functions, some of which are very specialized, is beyond the scope of this book.

Let us consider two functions in order to familiarize ourselves with these file interfaces. Below we shall consider in more detail the functions `xlsread` and `xlswrite` for communication with Microsoft Excel.

The file `ExcelDatEx.xls`, which we have already used in Problem 7 in connection with the array editor contains an Excel sheet ("Table 1") with a numerical field (matrix). This is the simplest form of Excel file, which can be processed by `xlsread`. If the file is in an access path for MATLAB (in the simplest case, in the **current directory**), then this matrix can be read into the MATLAB workspace using the instruction

```
>> ExcelMat = xlsread('ExcelDatEx')

ExcelMat =

    3    1    2
    4    0   -5
    3    1    2
    4    0   -5
    3    1    2
    ...
```

and becomes available there under the assigned variable name (`ExcelMat`).

The following instructions can be used to convert a matrix into an Excel file:

```
>> fromcol1 = ExcelMat(1:5,1)

fromcol1 =

    3
    4
    3
    4
    3

>> xlswrite('excerptExcelDatEx', fromcol1, 'Part 1')
```

```
Warning: Added specified worksheet.
> In xlswrite>activate_sheet at 259
  In xlswrite at 213
```

The vector `fromcol1` will be written in the Excel file **excerptExcel-DatEx.xls**. In this way, as desired, a new Excel sheet with the name ("Part 1") is set up. A corresponding warning that this sheet has been newly set up is also given.

Numerical values can be exchanged between MATLAB and Excel by this method. If the Excel sheet contains other information (such as text), the value `NaN` (not a number) will be returned except in the case of caption text. That will be ignored.

The functions are equipped with numerous other options. For example, the data to be imported in an Excel file can be selected interactively. It is left to the reader to explore these details.

Other input-output interfaces of MATLAB will be discussed in the problems. Note Problem 37 in particular. This problem deals with audio files.

1.2.9 The MATLAB Search Path

In connection with file operations it seems appropriate to clarify how MATLAB searches for and finds commands, programs, and files.

Up to now we have only made use of the current directory. All the programs and files saved there are automatically recognized by MATLAB. Beyond that, MATLAB must also have access to all the commands and function definitions that belong to the vocabulary of the MATLAB core and the toolboxes. The pertinent directories are saved in the MATLAB search path, which is read in on starting.

The current access paths of MATLAB can be listed as well as set using the *path* command.

The current search paths are listed by giving the instruction

```
>> path

   MATLABPATH

   C:\matlabR14\toolbox\matlab\ops
   C:\matlabR14\toolbox\matlab\lang
   C:\matlabR14\toolbox\matlab\elmat

   ...
```

```
C:\matlabR14\work
C:\beucher\Matlab7
C:\beucher\Book\Matlab\accompanyingsoftware
```

...

In the above (reduced display) example, the directories for the function groups of the already mentioned MATLAB operators (`ops`) or the elementary mathematical functions `elmat` are listed.

The directories `C:\beucher\...` are ones that the author himself has appended to the search path.

If you want to tie in a path, such as for the directory `C:\beucher\Matlab7`, you can do it very rapidly with the following command:

```
>> path(path, 'C:\beucher\Matlab7');
```

Of course, a directory of this sort can also be appended very simply (and then removed) by a menu command, in this case with `File - Set path ...`.

MATLAB automatically finds all the m-files (commands, specific MATLAB programs, Simulink systems) in the directories indicated by `path`.

If you write your own programs (see Section 1.6) it is recommended that you set up your own work directory in the search path. Programs must not, however, be put in any toolbox directory. During the next automatic update of this toolbox, these will be lost unless they are otherwise saved.

PROBLEMS

Work through the following problems on importing and exporting external files.

NOTE ▶ Solutions to all problems can be found in Chapter 4.

Problem 33

Generate a vector and/or a matrix of real numbers with an editor and save them under a file name.

Erase all the matrices in the work space using `clear`.

Then try to read in the file content using the `load` command and analyze the content of the workspace.

Be sure to compare this with the following Problem 34.

Problem 34
Generate an ASCII file with a column vector of complex numbers using a normal editor. Export this via the MATLAB command plane into the MATLAB workspace. What do you find?

Problem 35
Generate a column vector with MATLAB from complex numbers. Save this vector using **save** in binary format.

Next erase it in the workspace using the **clear** command and then reload the file into the MATLAB workspace using **load** or the import wizard.

Compare the result with that of Problem 34.

Problem 36
Define the complex matrix

$$M = \begin{pmatrix} 1 & 2 & 3 \\ 1+j & 6 & 0 \\ 3+j & 0 & -j \end{pmatrix}.$$

in the MATLAB workspace. Next, save this vector as a *column vector* in ASCII format as a file named **Mmatrix.txt** using the function **dlmwrite**.

Consult MATLAB help (**help dlmwrite**) about using this function.

Problem 37
Using a suitable MATLAB function read in one of the ***.wav** files to be found under **C:\WINDOWS\MEDIA** and display the corresponding audio signal graphically. Make sure that the time axis is labeled with the correct times.

Next, multiply the signal by a factor of 10 and save it in ***.wav** format under another name.

How do the comparative audio signals sound?

Problem 38
Solve Problem 37 again with "Microsoft Sound" (file **WindowsXP-Startup.wav**) and a factor of 0.

Problem 39
Prepare a folder **C:\mymatlab** and arrange it so that you can use it as a directory for MATLAB.

1.2.10 Elementary Matrix Manipulations

Among the most often used MATLAB commands are some of the so-called elementary matrix manipulations. The command **help elmat** provides a complete survey of these commands.

In earlier sections we have already used a couple of these commands, such as **zeros** and **ones** to create matrices of zeros or ones. These are very useful, mainly for initializing vectors or matrices.

Here, a 2×2 matrix with zeros and a 3×2 matrix with ones are created, and a vector with the length of the previously defined vector $\vec{x}_1$ is initialized to zeros.

```
>> M = zeros(2,2)

M =

        0       0
        0       0

>> N = ones(3,2)

N =

        1       1
        1       1
        1       1

>> x1 = [1,2,3,4,5,6];
>> v=zeros(length(x1),1)

v =

        0
        0
        0
        0
        0
        0
```

The command **eye** is often useful for generating a unit matrix. Here is an example:

```
>> E5 = eye(5)

E5 =

        1       0       0       0       0
        0       1       0       0       0
        0       0       1       0       0
```

```
    0    0    0    1    0
    0    0    0    0    1
```

The length of a vector and the size of a matrix can be determined automatically using the commands **length** and **size**. This is of great importance, especially in programs. Next the length of the just defined vector $\bar{x}_1$ is determined this way:

```
>> length(x1)

ans =

    6
```

The following sequence of commands eliminates the second column of the unit matrix **E5** created above and then determines the size of the resulting matrix:

```
>> B = E5;           % store the matrix E5
>> B(:,2) = [ ]      % empty the second column

B =

    1    0    0    0
    0    0    0    0
    0    1    0    0
    0    0    1    0
    0    0    0    1

>> [rows, columns] = size(B)    % determine sizes

rows =

    5

columns =

    4
```

Other commands for the generation of special matrices and special variables and constants are included in the commands listed in **help elmat** for MATLAB. We leave it to the reader to explore these commands, but we do not want to omit special mention of the command **why**.

The operator ' for *transposition* of a matrix or vector should be mentioned in connection with the elementary matrix operations, although this operator is not listed under the corresponding rubric.

A matrix is transposed when the ' symbol is placed after the matrix that is to be transposed.

For example, the following sequence of commands yields identical column vectors $\vec{x}_1$ and $\vec{x}_2$:

```
>> x1 = [1; 2; 3; -1; 4; 5]     % column vector definition

x1 =

        1
        2
        3
       -1
        4
        5


>> x2 = [1  2  3  -1  4  5]'   % row vector transpose

x2 =

        1
        2
        3
       -1
        4
        5
```

and the sequence of commands

```
>> M = [1 2; 3 -2; -1 4]          % a 3x2 matrix

M =

        1     2
        3    -2
       -1     4

>> N = M'
```

```
N =

    1      3     -1
    2     -2      4
```

operating on the 3×2 matrix

$$M = \begin{pmatrix} 1 & 2 \\ 3 & -2 \\ -1 & 4 \end{pmatrix}$$

generates the transposed matrix

$$N = \begin{pmatrix} 1 & 3 & -1 \\ 2 & -2 & 4 \end{pmatrix}.$$

In accordance with the rules of mathematics, complex matrix terms subjected to the operator ' will be folded and their conjugates taken:

```
>> M = [i 2; 3 -j]        % a 2x2 matrix with complex terms

M =

            0 + 1.0000i   2.0000
        3.0000                0 - 1.0000i

>> N = M'                  % the transposed matrix
N =

            0 - 1.0000i   3.0000
        2.0000                0 + 1.0000i
```

Thus,

$$M = \begin{pmatrix} j & 2 \\ 3 & -j \end{pmatrix}$$

yields

$$N = \begin{pmatrix} -j & 3 \\ 2 & j \end{pmatrix}$$

also.

If taking the conjugate of the terms is to be avoided, then the transposition operator has to be redefined as a field operator:

```
>> K = M.'
```

```
K =
```

$$
\begin{array}{lr}
0 + 1.0000i & 3.0000 \\
2.0000 & 0 - 1.0000i
\end{array}
$$

Then, in this case,

$$ M = \begin{pmatrix} j & 2 \\ 3 & -j \end{pmatrix} $$

yields

$$ K = \begin{pmatrix} j & 3 \\ 2 & -j \end{pmatrix} $$

as desired.

We have already learned about the **:** operator in Section 1.2.1 in connection with access to columns and rows in a matrix. The following application of this operator, with which a column vector with the same terms can be created from a matrix, is also very useful:

```
>> M = [1 2; 3 -2; -1 4]       % a 3x2 matrix

M =

        1     2
        3    -2
       -1     4

>> mVec = M(:)

mVec =

        1
        3
       -1
        2
       -2
        4
```

In this way the columns of the matrix are organized one above the other.

The function **repmat** can be used to copy and duplicate matrix entries. The instruction

```
>> N = repmat(M,2,2)
```

```
N =

    1     2     1     2
    3    -2     3    -2
   -1     4    -1     4
    1     2     1     2
    3    -2     3    -2
   -1     4    -1     4
```

reproduces the above matrix M four times in a 2×2 arrangement and saves it as the variable N.

In conclusion, we mention the **end** operator, which is especially useful in MATLAB programming. This operator evaluates the last index of a vector. The following sequence of instructions can thus be used to erase the last term of a vector without explicitly specifying the index:

```
>> zVec = [0, 3, -1, 0, 1, 99]

zVec =

    0     3    -1     0     1    99

>> zVec(end) = [ ]

zVec =

    0     3    -1     0     1
```

This technique can be used especially within programs (see Section 1.6) when the size of a vector changes during execution of a program so that the last index is not known explicitly during programming. A vector can also be lengthened in this way, as the following instructions show:

```
>> yVec = [0, 3, -1, 0, 1, 99]

yVec =

    0     3    -1     0     1    99

>> yVec(end+1:end+3) = [-1 -2 -3]
```

```
yVec =

     0     3    -1     0     1    99    -1    -2    -3

>> yVec(end+1) = 100

yVec =

     0     3    -1     0     1    99    -1    -2    -3   100
```

PROBLEMS

Work through the following problems on elementary matrix manipulations. Where needed, consult `help elmat` in the MATLAB help in order to find suitable functions.

NOTE ➤ Solutions to all problems can be found in Chapter 4.

Problem 40
Define the vector

$$\vec{r} = \begin{pmatrix} j & j+1 & j-7 & j+1 & -3 \end{pmatrix}$$

in MATLAB as a column and row vector.

Problem 41
Redo Problem 2, using the function `repmat`.

Problem 42
Using a suitable elementary matrix manipulation, generate a vector that has ten points between 0 and 1 with equal relative distances between them on a *base-10 logarithmic scale*. This means that the *base-10 logarithms* of the generated numbers should be separated by equal distances.

Next, plot these points using `loglog` as a check.
What sort of graph should you see?

Problem 43
Generate the graphs in the three-dimensional plot example in Section 1.2.5 using the function `meshgrid` to create the grid pattern.

Problem 44

In what way can you define the vector

$$\vec{y} = (1, \ 1.1, \ 1.2, \ 1.3, \ 1.4, \ \cdots, \ 9.8, \ 9.9, \ 10)$$

in MATLAB?

Problem 45

Using a suitable elementary matrix manipulation, reverse the sequence of terms in the vector of Problem 44; that is, create the vector

$$\vec{y} = (10, \ 9.9, \ 9.8, \ 9.7, \ \cdots, \ 1.2, \ 1.1, \ 1).$$

Problem 46

Consider the vector

$$\vec{z} = (1, \ 1.5, \ 2, \ \cdots, \ 98.5, \ 99, \ 99.5, \ 100).$$

Using logical operations, construct a vector $\vec{w}$ that contains every third component of $\vec{z}$.

Hint: use `repmat`.

1.3   MORE COMPLICATED DATA STRUCTURES

It has already been explained in Section 1.1 that in the newer versions of MATLAB, as in other higher programming languages, additional data structures besides classical numerical matrices are available.

In the following sections we discuss so-called *structures* and *cell arrays* in more detail.

These data structures are seldom called upon in the interactive mode. On the other hand, in MATLAB programming they allow constructions (see Section 1.6), which would be very intricate or entirely unrealizable using numerical fields. Thus, the full value of these data structures will only become clear to the reader after reading Section 1.6. In the following we shall just discuss the basic aspects of these constructions in the interactive mode.

1.3.1   Structures

Structures are fields (arrays) by which different types of data can be combined into a logical unit. Access to these data is obtained by *names*, rather than by numerical indexing as in the case of matrices.

Definition of Structures

We illustrate this with an example. Suppose we want to combine certain properties of a MATLAB graphic, such as

- the title of the graphic,
- the label on the x axis,
- the label on the y axis,
- the number of plots,
- the colors of the plots,
- whether a grid is set (yes=1/no=0),
- the x range to be displayed, and
- the y range to be displayed.

This listing includes numbers, numerical vectors, and strings. A listing of this sort can, of course, no longer be represented by a matrix in MATLAB.

For this purpose, we define a structure **Graphic** in the following way:

```
>> Graphic.title = 'Example';
>> Graphic.xlabel = 'time / s';
>> Graphic.ylabel = 'voltage / V';
>> Graphic.num = 2;
>> Graphic.color = ['r', 'b'];
>> Graphic.grid = 1;
>> Graphic.xVals = [0,5];
>> Graphic.yVals = [-1,1];
>> whos
   Name            Size                    Bytes  Class

   Graphic         1x1                      1096  struct array

Grand total is 42 elements using 1096 bytes
```

Evidently, the above instructions have been used to define a *single* element (1 × 1 array) **Graphic**, which contains the full information, as is shown by a call of **Graphic**:

```
>> Graphic

Graphic =

    title: 'Example'
   xlabel: 'time / s'
   ylabel: 'voltage / V'
```

```
     num: 2
   color: 'rb'
    grid: 1
   xVals: [0 5]
   yVals: [-1 1]
```

Alternatively, a structure can also be defined using the MATLAB function **struct**. For example, a variable of the above graphic structure could be pre-initialized with empty fields as follows:

```
>> Graphicempty = struct('title', [], 'xlabel', [], ...
                         'ylabel', [], 'num', [], ...
                         'color', [], 'grid', [], ...
                         'xVals', [], 'yVals', [])

Graphicempty =

   title: []
  xlabel: []
  ylabel: []
     num: []
   color: []
    grid: []
   xVals: []
   yVals: []
```

Fields of structures (*structure arrays*) can also be set up. The following instructions set up a 10×1 field of empty graphic structures:

```
>> Grfarray = repmat(Graphicempty, 10, 1)

Grfarray =

10x1 struct array with fields:
    title
    xlabel
    ylabel
    num
    color
    grid
    xVals
    yVals
```

The content of the third component of this field is then

```
>> Thirdarray = Grfarray(3)
```

```
Thirdarray =

        title: []
       xlabel: []
       ylabel: []
          num: []
        color: []
         grid: []
        xVals: []
        yVals: []

>> whos
  Name              Size                Bytes  Class

  Graphic           1x1                  1096  struct array
  Graphicempty      1x1                   992  struct array
  Grfarray          10x1                 5312  struct array
  Thirdarray        1x1                   992  struct array

Grand total is 138 elements using 8392 bytes
```

Here you can see that indexing occurs as with numerical fields.

As noted above, the major significance of combining data into a structure will first become clear to the reader in connection with programming of MATLAB functions. The logical combination of individual data into a unit makes programming easier, especially for extensive tasks involving enormous programs, and makes the programs more readable. This applies particularly to the transfer of parameters. Long, involved parameter lists can be shortened by using structures and the programs can made more transparent for both the programmer and the user (see the solution to Problem 58).

At this point we cannot, of course, pursue this aspect further. We conclude with a discussion of elementary operations, such as the definition of structures and the operations for access to structure arrays (structure fields).

Access to Structure Arrays

Access to the data fields of structures is obtained by *specifying the name* of the structure and the desired field. Thus, for example, with

```
>> ttlString = Graphic.title

ttlString =

Example
```

one can obtain access to the data field `title` of the structure `Graphic` and assign the result to the variable `ttlString`.

In similar fashion a new value can be assigned to the data field:

```
>> Graphic.title = 'Another example'

Graphic =

    title: 'Another example'
   xlabel: 'time / s'
   ylabel: 'voltage / V'
      num: 2
    color: 'rb'
     grid: 1
    xVals: [0 5]
    yVals: [-1 1]
```

The same could also be obtained using the MATLAB function `setfield`:

```
>> Graphic = setfield(Graphic, 'title', 'The next example')

Graphic =

    title: 'The next example'
   xlabel: 'time / s'
   ylabel: 'voltage / V'
      num: 2
    color: 'rb'
     grid: 1
    xVals: [0 5]
    yVals: [-1 1]
```

If access to elements in structure arrays is desired, then the field components must be addressed by numerical index and the data field by name. For example, the instructions

```
>> Grfarray(3).title = 'Third graphic';
>> Grfarray(3).xVals = [0,10];
>> Grfarray(3)

ans =

    title: 'Third graphic'
   xlabel: []
   ylabel: []
```

```
      num: []
    color: []
     grid: []
    xVals: [0 10]
    yVals: []
```

set up the data array `title` for the third structure in the array (field) `Grfarray` as the string `'Third graphic'` and the data field `xVals` for the third structure of the array `Grfarray` on the interval (numerical field) `[0 10]`.

The next example concerns numerical indexing to a field component *and* to a data field. The upper bound of the range of the graph is to be changed and written into the corresponding structure. This can be done via the following instruction:

```
>> Grfarray(3).xVals(2) = 25;
>> Grfarray(3)

ans =

     title: 'Third graphic'
    xlabel: []
    ylabel: []
       num: []
     color: []
      grid: []
     xVals: [0 25]
     yVals: []
```

The function `deal` is also of practical value for dealing with structure arrays. A data field can be filled for *all* components of a structure array using this function. The instruction

```
>> [Grfarray.xlabel] = deal('voltage / V');
>> Grfarray(1)

ans =

     title: []
    xlabel: 'voltage / V'
    ylabel: []
       num: []
     color: []
      grid: []
```

```
        xVals: []
        yVals: []

>> Grfarray(7)

ans =

        title: []
       xlabel: 'voltage / V'
       ylabel: []
          num: []
        color: []
         grid: []
        xVals: []
        yVals: []
```

sets all the `xlabel` data fields in the structure array `Grfarray` to the string `'voltage / V'`.

Modifying a Structure

A structure can be modified simply by removing fields or by appending fields. Thus, for example,

```
>> Graphic = rmfield(Graphic, 'color')

Graphic =

        title: 'The next example'
       xlabel: 'time / s'
       ylabel: 'voltage / V'
          num: 2
         grid: 1
        xVals: [0 5]
        yVals: [-1 1]
```

causes the field `color` to be eliminated from the structure. We could, for example, then replace it with a structure in which color *and* line style are combined:

```
>> Style.color = 'rb';
>> Style.line = '-o';
>> Graphic.style = Style
```

```
Graphic =

      title: 'The next example'
     xlabel: 'time / s'
     ylabel: 'voltage / V'
        num: 2
       grid: 1
      xVals: [0 5]
      yVals: [-1 1]
      style: [1x1 struct]
```

Here, as well, the MATLAB function **setfield** can be used:

```
>> Graphic = setfield(Graphic, 'NewDataField', 5)

Graphic =

          title: 'The next example'
         xlabel: 'time / s'
         ylabel: 'voltage / V'
            num: 2
           grid: 1
          xVals: [0 5]
          yVals: [-1 1]
          style: [1x1 struct]
    NewDataField: 5
```

This example also shows that structures, themselves, can be data fields of structures. Likewise, as shown above, structures can, in turn, be combined into arrays.

PROBLEMS

> NOTE   Solutions to all problems can be found in Chapter 4.

Problem 47
Modify the field **yVals** of the structure **Graphic** so that it has the value [-2 0]. See if there are different ways of doing this.

Problem 48
Save the second line color that was saved in the **style** partial structure as a variable of its own.

Problem 49

Define a structure `color` with the data fields `red`, `blue`, and `green`.

Then define a 1×20 field of structures of this type and initialize the `red` component with the value `'yes'`, the `blue` component with the value `'no'`, and the `green` component with the value `[0,256,0]`.

Problem 50

Eliminate everywhere the data field `blue` from the structures of the structure field defined in Problem 49 and set the value of `green` of the tenth structure to `[0,256,256]`.

Problem 51

An attempt to eliminate the entry `blue` in the structure field of Problem 49 *in only the first field component* fails. There is no operation or MATLAB function that can do this.

Why is that?

1.3.2 Cell Arrays

As with structures, data of different types can be combined into a unit with so-called *cell arrays*. The data field will then no longer be addressed by names, but can, like numerical fields, be reached through *numerical indexing*.

At the same time, there must be a distinction between the (indexed) access to the components of the cell array itself and access to the content of the cells. Hence, two kinds of indexing are used with cell arrays: *cell indexing* and *content indexing*. These will be recognized by the brackets used for indexing. Curly brackets `{}` are used for content indexing and parentheses `()` for cell indexing. With these different modes of access, cell arrays are widely used.[13]

In the following we shall try to make this somewhat cryptic introduction more understandable with some examples.

Let us suppose that we are measuring the temperature in the ocean at many points and at different depths. We can specify the position of a measurement by (x, y) coordinates. The measurements will be saved in the form (`depth`, `temperature`), with the number of measurements at a given position not set in advance. An entire measurement campaign is to be saved in this data record. It should be possible to gather the measurements from a week into a data structure.

---

[13]Readers who do not feel confident enough, despite the basic preparation for MAT-LAB in the first section of this book, can defer the following section until the concept of cell arrays is discussed in Section 1.6.

This problem can perhaps be solved using structures, by setting up structure arrays of the form

```
measurement = struct('day', [], 'x', [], ...
                      'y', [], 'depth', [], ...
                      'temperature', [], 'measurementnumber', [])

measurement =

                 day: []
                   x: []
                   y: []
               depth: []
         temperature: []
   measurementnumber: []
```

If you want to search in a program for the 25th measurement on the 3rd day, corresponding search loops must be written (see Section 1.6.3), which search the entire data record.

In this case it might be simpler to index with a two-dimensional field (e.g., **meascampaign(3,25)**, in which the first component is the day and the second, the series number of the measurement). The data cannot be written in the form of a numerical field, even though they are numbers, because the data records are of varying lengths.

This can, however, be done with cell arrays.

1.3.3 Definition of Cell Arrays

In order to define the desired data structures in a step-by-step manner, let us first simplify the problem and consider the data that could be acquired for a *single measurement*. These include the coordinates—a two-dimensional vector—and the depth/temperature data, which for n measurements yields an $n \times 2$ matrix. Of course, we *cannot* (meaningfully) combine these data into a single unit in the form of a numerical field. In the case $n = 3$ the following instructions obviously lead to an error message, since the numerical fields have different dimensions:

```
>> measurement = [256.9, 300.7]        % coordinates

measurement =

   256.9000   300.7000
```

```
>> % measurement data
>> measurement = [measurement, [10, 27; 50, 16; 100, 5]]
??? Error using ==> horzcat
All matrices on a row in the bracketed expression must have the
 same number of rows.
```

The desired logical combination can be realized easily using the following instructions:

```
>> clear
>> measurements = {[256.9, 300.7], ...
               [10, 27; 50, 16; 100, 5]}

measurements =

    [1x2 double]    [3x2 double]

>> whos
  Name             Size           Bytes  Class

  measurements     1x2              184  cell array

Grand total is 10 elements using 184 bytes
```

As the command **whos** shows, the variables **measurements** now form a *two-dimensional cell array* consisting of a 1×2 vector (the coordinates) and a 3×2 matrix (the measurements taken at this position). The curly brackets { } on the right-hand side of the variable assignment indicate that a cell array has been defined here.

At this point the instruction

```
>> clear measurements
```

should be entered. This instruction erases the previously defined numerical vector with the same name. This must be done in order to keep the subsequent definition from generating an error message. Essentially, in the case of cell arrays, preexisting definitions are *not overwritten* by a new definition. They must first be erased.

Alternatively, the cell array **measurements** could also have been defined in the following way:

```
>> clear measurements
>> measurements(1) = {[256.9, 300.7]}
```

```
measurements =

    [1x2 double]

>> measurements(2) = {[10, 27; 50, 16; 100, 5]}

measurements =

    [1x2 double]    [3x2 double]
```

Here, the different significance of the two types of brackets, {} and (), mentioned previously, should be kept in mind. According to the above instructions, the cells in the cell arrays **measurements** will be indexed on the left of the assignment using *parentheses*. This is a matter of the *cell indexing* noted above. On the right, the assignment will indicate by {} brackets around the object to be saved that a cell array is being defined.

A definition that uses *content indexing* follows this syntax:

```
>> clear measurements
>> measurements{1} = [256.9, 300.7]

measurements =

    [1x2 double]

>> measurements{2} = [10, 27; 50, 16; 100, 5]

measurements =

    [1x2 double]    [3x2 double]
```

Now the cells of the cell array **measurements** are indexed on the left-hand side of the instruction using *curly brackets*. In this way it is also made clear that a cell array is being defined. On the right-hand side of the assignment, the object to be stored is specified (without curly brackets or parentheses).

The cell array **measurements** defined in this way can be visualized graphically using the function **cellplot**. The instruction

```
>> cellplot(measurements)
```

generates the graphic shown in Fig. 1.19.

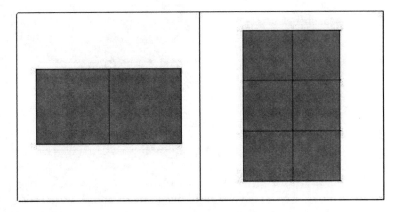

FIGURE 1.19   **Visualization of the cell array** measurements **with** cellplot.

This graphic shows the individual cells of the cell array in terms of their field dimension, not their content. These can be shown in a single stroke using the function **celldisp**:

```
>> celldisp(measurements)

measurements{1} =

  256.9000  300.7000



measurements{2} =

     10      27
     50      16
    100       5
```

We can now define the originally conceived cell array **meascampaign** using cell arrays of the form **measurements** :

A field of this sort can be built using the instructions

```
>> meascampaign{1,1} = measurements;
>> meascampaign{1,2} = { [122.7, 103.1], ...
                 [5, 29; 50, 13; 100, 4; 200, 2]};
>> meascampaign{3,5} = { [101.0, 200.0], ...
                 [5, 31; 50, 22; 100, 12; 150, 6; 200, 1]}
```

```
meascampaign =

  {1x2 cell}    {1x2 cell}     []     []           []
       []            []        []     []           []
       []            []        []     []     {1x2 cell}
```

As you can see, in this example the first two measurements on day 1 will be stored and then the 5th measurement on day 3. The fields in between will be automatically created and filled with the empty field [ ]. The individual elements in the cells are evidently, *in turn*, cell arrays. Calling `cellplot`

```
>> cellplot(meascampaign)
```

generates the overview of the structure of **meascampaign** shown in Fig. 1.20.

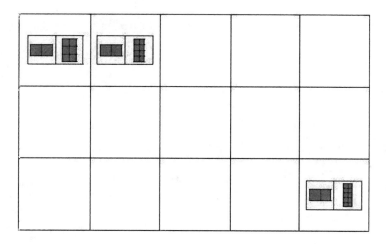

FIGURE 1.20   **Visualization of the cell array** measurements **with** cellplot.

1.3.4 Access to Cell Array Elements

Just as with the definition of cell arrays, *access* to the data in a cell array demands that attention be paid to the two different indexing schemes, *cell indexing* and *content indexing*.

In the above example, *content indexing* ({ }) was used to gain access to the contents of the cells in the cell array **measurements** in order to store them in numerical variables:

```
>> position = measurements{1}
```

```
position =

  256.9000   300.7000

>> value = measurements{2}

value =

      10     27
      50     16
     100      5
```

On the other hand, with

```
>> cell1 = measurements(1)

cell1 =

   [1x2 double]

>> cell2 = measurements(2)

cell2 =

   [3x2 double]
```

cell indexing is used to gain access to the cells themselves (!), in order to store them as cells (!) in the new cell arrays **cell1** and **cell2**, as a glance at the workspace shows:

```
>> whos
   Name          Size              Bytes  Class

   cell1         1x1                  76  cell array
   cell2         1x1                 108  cell array
   meascampaign  3x5                 828  cell array
   measurements  1x2                 184  cell array
   position      1x2                  16  double array
   value         3x2                  48  double array

Grand total is 79 elements using 1260 bytes
```

To get access to the data in the cell array **meascampaign** we have to gain access step-by-step with *multiple indexing*, since the cells of

meascampaign themselves contain cell arrays which, in turn, contain numerical arrays:

```
>> meas1day1 = meascampaign{1,1}

meas1day1 =

    [1x2 double]    [3x2 double]

>> iscell(meas1day1) %test: is it a cell array?

ans =

    1

>> position1day1 = meascampaign{1,1}{1}

position1day1 =

  256.9000   300.7000

>> measpos1day1 = meascampaign{1,1}{2}

measpos1day1 =

     10     27
     50     16
    100      5

>> meas3pos1day1 = meascampaign{1,1}{2}(3,:)

meas3pos1day1 =

    100      5

>> tempmeas3pos1day1 = meascampaign{1,1}{2}(3,2)

tempmeas3pos1day1 =

    5
```

In order to understand what happens, let us take a look at the last instruction as an example. With **meascampaign{1,1}** the *contents* of the cell with the coordinates $(1, 1)$ are reached. This is, again, a cell array, whose second cell component is reached with **{2}**. This second cell component contains, in turn, a 3×2 matrix. The second component in the third row of this matrix is specified by **(3,2)**, in the conventional indexing for numerical fields.

Modifying Cell Arrays

As already noted, caution is called for when redefining cell arrays if a variable with the same name already exists. For example,

```
>> measurements = [200, 30]

measurements =

    200    30
```

yields no error message—the existing cell array will simply be overwritten—but

```
>> measurements{1} = [256.9, 300.7]
??? Cell contents assignment to a non-cell array object.
```

yields an error message, since the (numerical) field **measurements** already existed.

Individual components of a cell array can, of course, also be changed using the mechanisms described above. Thus, for example, with

```
>> meascampaign{1,1}{2}(3,:) = []

meascampaign =

    {1x2 cell}    {1x2 cell}    []    []        []
           []            []    []    []        []
           []            []    []    []   {1x2 cell}
```

the last measurement on the 5th day will erased; but this cannot be recognized initially in the MATLAB answer since the whole cell array is returned. The instruction

```
>> celldisp(meascampaign)

meascampaign{1,1}{1} =

    256.9000   300.7000
```

```
meascampaign{1,1}{2} =

      10      27
      50      16


      ...
```

shows, however, that the originally defined row **100, 5** from **meascampaign{1,1}** is absent.

PROBLEMS

Work through the following problems on the topic of cell arrays.

NOTE Solutions to all problems can be found in Chapter 4.

Problem 52
Change all the measured values to [-1 , 0] in the cell array **meascampaign** in the cell with coordinates (3, 5).

Problem 53
Extend the cell array **measurements** by one cell in which the name of the engineer who made the measurement is entered.

Then generate a 5 × 1 cell array **weeklymeasurements**. Fill each cell with the cell array **measurements**. Then, for each day of the week enter an (arbitrary) name in the prescribed location.

Problem 54
Make the following changes in the cell array **weeklymeasurements** defined in Problem 53:

1. Change the name of the engineer who made the measurement on the 5th day to **A. E. Neumann**.
2. Add a measurement with the values **250,0** to the 1st day.
3. Save the data from the 3rd day in a separate cell array **day3**.
4. Using the function **deal**, whose syntax can be found under help, save the measurements for all five days in separate variables, **monday**, **tuesday**, **wednesday**, **thursday**, and **friday**.

The next problem introduces another useful I/O function, **textscan** (see Section 1.2.6). It is noteworthy in that it makes use of cell arrays. The command `textscan` can read formatted data from a text file opened with the function `fopen` and arranges the results of the reading operation in a cell array.

Familiarize yourself with the syntax of these functions and then work out the following.

Problem 55

 Open the file `sayings.txt` in the accompanying software using `fopen` and read the first seven strings, which are separated by spaces, using the function `textscan` into a cell array `sayings`. Then close the file using `fclose`. Next, display the result of this reading operation on the screen using `celldisp` and `cellplot`.

1.4 THE MATLAB DESKTOP

With MATLAB 7 the MATLAB user interface has again undergone extensive modernization, which really makes using this tool substantially easier to use, both in the interactive mode and for developing programs (see Section 1.7). In this section these capabilities will be elucidated further. Of course, we cannot give the reader a complete overview of the manifold capabilities of the desktop. That would go beyond the scope of this book.

Just a few of the basic principles will be discussed.

Menus and Shortcuts

Basically, the most varied capabilities involve addressing or setting up the MATLAB user interface (see Fig.1.1).

In general, the functions of the user interface can be selected by

- choosing the appropriate menu entry using the mouse,
- choosing the appropriate icons in the icon toolbar using the mouse, so-called *shortcuts*, and
- selection within a *context menu* (i.e., by right-clicking the mouse in the current working window).

Beyond this, many functions can also be called by entering appropriate MATLAB commands in the command window.

The first two possibilities for choosing functions are standard in every Windows application and need not be elaborated further here.

As for *shortcuts*, a distinction must first be made between the so-called *keyboard shortcuts* and *MATLAB shortcuts*. *Keyboard shortcuts* are standard in every Windows application. They are key combinations that set up a given functionality in the user interface. The best known and most used representatives of this class are `Control-C` and `Control-V` for copying and pasting data between two Windows applications with the aid of the Windows interface. In the MATLAB command window, for example, the key combinations `Control-P` and `Control-N` replace the arrow keys ↑ and ↓ for reconstructing commands, as mentioned in Section 1.2.1.

With *MATLAB shortcuts*, on the other hand, MATLAB commands can be combined and later called in a single step with a mouse movement. Shortcuts are defined using the so-called *shortcut editor*, which can be called up with the *start button* (see Fig. 1.1, item 7) in the menu `Shortcuts - new shortcut...`. Based on an option chosen within the editor, the defined shortcuts can either be anchored in the start-button menu or integrated as icons in the shortcut toolbar (item 6).

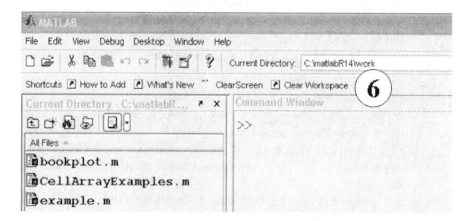

FIGURE 1.21 **Shortcut toolbar with the self-defined shortcut "clear screen."**

Fig. 1.21 shows a segment of Fig. 1.1 with an icon integrated into the shortcut toolbar for initiating the often-used MATLAB command `clc`, with which the command window can be cleared.

Likewise, a fast way of selecting functionalities in the MATLAB interface is to right-click the mouse on the available *context menu* in the corresponding working window (context!).

Right-clicking the mouse then opens a context related pulldown menu in which functions can be selected by left-clicking the mouse.

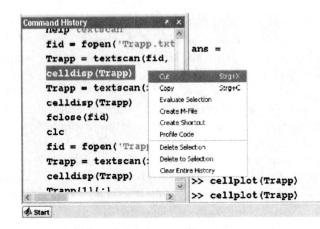

FIGURE 1.22   Context menu in the command-history window.

As an example, Fig. 1.22 shows a pulldown menu generated by right-clicking the mouse in the command-history window.

In this menu it is now possible to select what is to be done with it or with the previously selected commands in the history window. For example, they can be executed again by choosing the menu item `Evaluate Selection`.

Tab Completion

An interesting possibility for making MATLAB easier to use in the interactive mode is the *tab completion* technique. *Variable names*, *function names*, *file names*, and *structures*, as well as *graphic handles*, can be quickly and efficiently created (in the command window) using the tab key on the PC keyboard.

To do this, the functionality is first initiated by choosing the appropriate box in the menu `File- Preferences ...` under `Command-Window- Keyboard ...`.

Then, for example, you can type in the beginning of the function name in the command window and press the tab key.

Fig. 1.23 shows the result of this operation after entering `plo`, the first letters of many plot functions.

The desired function can be selected in the pulldown menu. This is then entered in the command window. The prerequisite for operation of this mechanism is, of course, that MATLAB can also find the corresponding function or variable names (i.e., that these names are entered in the current search path or in the current workspace).

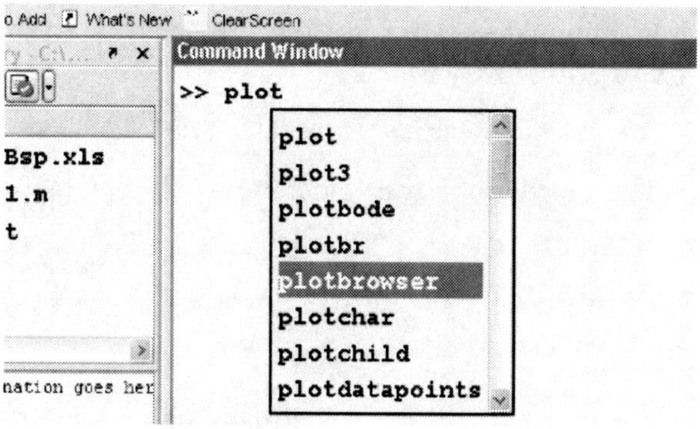

FIGURE 1.23 **Tab-completion function.**

History and Monitoring Functions

The history mechanism, which was already introduced in Section 1.2.1, is of special interest for the MATLAB user on returning to the interactive mode after a long time. The history functionality allows the user to reconstruct commands from the current session or from earlier sessions and to reuse them rapidly.

All the commands set in the course of the session are listed in the command-history window. These commands can then be reconstructed as shown in Section 1.2.1.

Beyond this, shortcuts (see above) or m-files (see Section 1.6) can easily be created from the commands collected in the command-history window. This is most easily done by selecting the relevant commands and by choosing the corresponding function in the context menu (right mouse click).

The search for old commands has been further simplified in MATLAB 7. Instead of searching in the command window, the reader can also enter the initial letter of a command in the command-history window. (Click on the window and type in the letters.) The window jumps to the next command with this initial letter. Whole parts of commands can also be found via the menu entry **Edit - Find** in the user interface.

For keeping a record of the entire session, not just the commands but also the responses of MATLAB, the **diary** function is available. This is chosen by entering

```
>> diary(filename)
```

From this point in time a record will be kept in ASCII format of the MATLAB session in a file with this name. The file is created in the current directory.

For example, if the session is saved in a file named `session 11.09.2005.txt` with

```
>> diary ('session 11.09.2005.txt')
```

it will be saved until this mechanism is terminated with the following command:

```
>> diary off
```

Configuration Capabilities

In Sections 1.2 and 1.2.1 it has already been pointed out that the interface shown in Fig. 1.1 does not in the least have a unified structure.

In fact, the user has many options available for molding the appearance of the interface according to his needs. The *docking mechanism* has already been described in Section 1.2.1. With it, partial windows in the interface can be docked and again be erased. Of course, the layout can also be varied by pulling with the mouse. Color arrangement, font, and many other aspects can be controlled via `Preferences`. Once the layout is changed, it can be saved using the menu entry `Desktop - Save layout ....`

It is left to the reader to create a suitable appearance for the layout of the MATLAB working interface.

1.5   MATLAB HELP

The help command `help` has often been referred to in the preceding sections. If the name of a command, of a group of commands, or of an entire toolbox is known, then the statement `help <function\_name>` can be entered in the command window to display the desired information in the command window.

This information can also be obtained by selecting the menu command `help - MATLAB help` and entering the corresponding command or command group in the search window under the `index` or `search` tab. This is especially useful when the precise name of the command is not known, since the search function permits a search for keywords in the full text of the MATLAB handbooks. At the same time, logical operations are possible. Search terms can be joined by the conjunctions **AND**, **OR**, and **NOT**.

Fig. 1.24 shows the help window after entry of `graphics AND axis AND label NOT properties` in the field `search for:` of the search tab.

Obviously the label functions `xlabel`, `ylabel`, and `zlabel` for MATLAB graphics introduced in Section 1.2.5 will be found with this search.

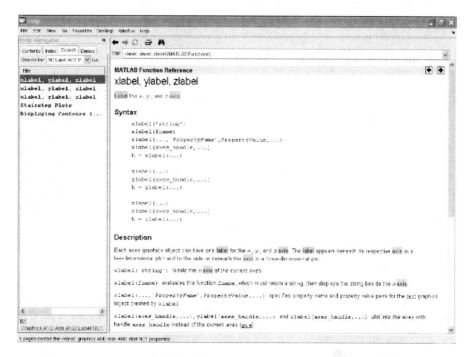

FIGURE 1.24 **The MATLAB help window after a search operation.**

Also, in the case of a keyword search, there is, as usual, another possibility in the form of a MATLAB command, which should not be omitted here. With the command

```
>> lookfor 'xlabel'
TEXLABEL Produces the TeX format from a character string.
XLABEL X-axis label.
IMTEXT Place possibly multi-line text as xlabel.
XLABEL4AX Set the XLABEL on a specified axis.
WSETXLAB Plot xlabel.
WXLABEL X-axis label.
```

all the help texts of the MATLAB command that lie in the search path after the keyword `xlabel` will be looked through. MATLAB functions can also be found very rapidly without leaving the command window by using the function `lookfor`.

In conclusion, it should be noted that the MATLAB documentation handbooks are also available in **pdf** format. This, of course, presumes that the pdf documentation is installed along with MATLAB and its toolboxes.

1.6   MATLAB PROGRAMMING

As noted at the beginning of this chapter, MATLAB is not only a numerical program for evaluating formulas, but is a *programming language* in itself. This means that, besides the commands mentioned previously, MATLAB includes such program language constructs as loops and branches and provides for the writing of functions and procedures.

These capabilities are discussed in this section.

1.6.1   MATLAB Procedures

The first step in programming is to provide *script files*, in which MATLAB command sequences are collected into simple procedures.

This involves writing the MATLAB command sequences with a text editor that can be opened, e.g., with the menu entry **File - new - m-file** in the MATLAB workspace (see Section 1.7), into a file and then saving this file under a name (e.g., **proced.m**) or as a so-called *m-file* in a path accessible to MATLAB. (It might be a good idea to check the execution of the **path** command in Section 1.2.6 again.)

The sequence of commands can then be executed in the command window with the command **proced**. In order to avoid problems, whenever possible you must not use any previously employed command name. You can check whether such a command name already exists by, for example, just typing **help <name>** in the command window. An even more reliable method is to use the function **exist**. If the call **exist <name>** returns a value of 0, then no function with that name exists in the search path.

We now illustrate the procedure for writing script files with an example. To do this, we have slightly expanded a sequence from Section 1.2.5,

```
>> t=(0:0.01:2);
>> sinfct=sin(2*pi*5*t);
>> cosfct=2*cos(2*pi*3*t);
>> expfct=exp(-2*t);
>> plot(t,[sinfct; cosfct; expfct])
```

and written an m-file with the name `fnExample.m`:

```
% Procedure fnExample
%
% call: fnExample
%
% first example of a script file
%
% book: MATLAB and Simulink
%
% Author: Prof. Dr. Ottmar Beucher
% Karlsruhe Polytechnic -- Engineering and Economics
%
% Version: 1.03
% Dates: 01/31/1998,3/21/2002,9/14/2005

t=(0:0.01:2);
sinfct=sin(2*pi*5*t);
cosfct=2*cos(2*pi*3*t);
expfct=exp(-2*t);
plot(t,[sinfct; cosfct; expfct])
xlabel ('time / s')
ylabel('Amplitude')
title('three gorgeous signals')
```

With

```
>> fnExample
```

this sequence can then be executed fully. At this point it is especially important to mention the heading of the file. With *% placed in front*, a documenting commentary can be provided ahead of the command sequence. The particular point here is not that the meaning of the program will still be understandable years later—documentation of this sort is part of good programming style and readers should get used to this early on in their own interest—but that this text can be read in the MATLAB command window by entering `help fnExample`. The help mechanism for all MATLAB functions, including for those you have written yourself, is set up so that `help <function name>` displays all the comment lines in the script or function file up to the first empty line or until the first MATLAB instruction. In order to save space, of course, we shall not fully reproduce all the comments for programs printed in this book. For this, refer to the accompanying software. The complete comments can, as noted above, also be displayed by entering `help <function name>` in the MATLAB command window.

Finally, it should be noted that, once the program is executed, the *variables* defined in the script file are *known in the MATLAB workspace*. This is an important distinction regarding the behavior of the MATLAB functions to be discussed below.

1.6.2   MATLAB Functions

It is far more flexible to define MATLAB *functions* rather than simply combine MATLAB commands in script files, since these will be able to pass on *parameters*.

With this mechanism a program can be executed under different conditions. This is most easily clarified with the aid of an example. To do this, we first modify the script file **fnExample.m** of the above example as follows:

```
function [t, sinfct, cosfct, expfct]= fnExample2(f1, f2, damp)
%
% Function fnExample2
%
% call:   [t, sinfct, cosfct, expfct] = fnExample2(f1, f2, damp)
% or      fnExample2(f1, f2, damp)
%
% First example of a MATLAB function
%
% Normally both a description of the input and output
% parameters and the important characteristics of the
% function are given here

t=(0:0.01:2);
sinfct=sin(2*pi*f1*t);
cosfct=2*cos(2*pi*f2*t);
expfct=exp(-damp*t);
plot(t,[sinfct;cosfct; expfct])
xlabel ('time / s')
ylabel('Amplitude')
title('three gorgeous signals')
```

and save it under the name **fnExample2.m**. Function names and file names must always be in agreement, since MATLAB searches for functions in an m-file *with the same name*. In particular, keep in mind that

MATLAB 7 distinguishes strictly between upper and lowercase letters[14] (*case sensitivity*).

What has changed in `fnExample2.m` compared to `fnExample.m`?

First of all, the three parameters `f1`, `f2`, and `damp` have been added. These correspond to the frequencies of the sine and cosine signals, plus a (damping) parameter for the exponential function.

These parameters, the *input parameters*, appear immediately after the function name as a list, separated by commas, in parentheses. Another list of names stands in front of the function name, this time in the square brackets typical of vectors. In the preceding example this list takes the vector names of the vectors used in the function core, which contain the results of the calculations. The equals sign in between signifies that we are dealing with *output parameters*. One should make sure that input and output parameters are described in detail in the comments preceding the actual instructions; this can be indicated with the help mechanism in MATLAB. This makes later use of the function much easier, especially for other users. Specifying the call syntax or giving an example of a call in the help commentary is also very practical. This can, for example, be copied directly into the command level after it is found with `help ...`, changed, if necessary, and then executed.

The following function call clarifies the significance of the difference between input and output parameters:

```
>> clear
>> [time, s1, c1, e1] = fnExample2(3, 5, 4);
>> whos
  Name        Size            Bytes  Class

    c1        1x201            1608  double array
    e1        1x201            1608  double array
    s1        1x201            1608  double array
    time      1x201            1608  double array

Grand total is 804 elements using 6432 bytes
```

You can tell that the names in the parameter list are merely *place holders* for the true values that will be set in the list when the function is called. The

---

[14]The reader will surely have noticed that the function names of original MATLAB commands in the help texts are always given in uppercase letters. This merely serves as emphasis in the help text. The commands, themselves, must be set in lowercase, as they are defined in that form.

variables are thus only known and valid within the function core and while the function is being executed. The variables are said to be *local*.

This is shown by calling **whos** after the function call.

Of course, variables rather than numbers can also be assigned to the function, and if one is no longer interested in the calculated vectors, but only in the graphic provided by the function for the case at hand, no return vector is needed. This is shown in the following example:

```
>> clear
>> whos
>> frq1=6;
>> frq2=1;
>> dmpng=7;
>> fnExample2(frq1,frq2,dmpng);
>> whos
  Name          Size              Bytes  Class

  ans           1x201              1608  double array
  dmpng         1x1                   8  double array
  frq1          1x1                   8  double array
  frq2          1x1                   8  double array

Grand total is 204 elements using 1632 bytes
```

As can be seen, the input parameters are now known (they were previously defined in the workspace) and the return vector is saved in the variable **ans**, which is customarily used as the return variable if no other has been specified.

Certainly, the parameters **f1**, **f2**, and **damp** are not known from the source text for the function. These are, as already noted, *place holders* for the actual delivered variables.

Of course, one should not be deceived by this example. The input parameters are, as before, *local*.

The following example makes this clear:

```
function [sum] = fnExample3(a,b)
%
% function fnExample3
%
% call: [sum] = fnExample3(a,b)
%
% second example of a MATLAB function
%
% Calculates the sum of two vectors a and b,
```

```
% which must, of course, also have the same dimensions,
% but which will not be checked in this version of the program.
%
% This should clarify the properties of local variables.

sum = a+b;       % It should do this.

                 % And now a little experiment.
a = a.^2;        % squaring the components of a
```

We now call these functions and see what happens:

```
>> clear
>> v1 = [1 2 3];
>> v2 = [-2 2 5];
>> thesum = fnExample3(v1, v2)

thesum =

    -1     4     8

>> whos
   Name         Size              Bytes  Class

   thesum       1x3                  24  double array
   v1           1x3                  24  double array
   v2           1x3                  24  double array

Grand total is 9 elements using 72 bytes

>> v1

v1 =

     1     2     3

>> v2

v2 =

    -2     2     5
```

As we can see, the function **fnExample3** calculates the sum of $\vec{v}_1$ and $\vec{v}_2$ in good form. These are still *not* changed here, although within the program the place holder of $\vec{v}_1$ is squared term by term. This is also done, but only with a local variable.[15] The input parameters themselves are protected.

In conclusion we should add that, as opposed to the script files, *within a function there is no* access to variables in the workspace. This is also a consequence of the concept of local variables. In addition, there is *also no call-by-reference mechanism*, as in the C/C++ programming language;[16] that is, results cannot be returned in the parameter list, but only through the return vector.

PROBLEMS

Work through the following problems in order to get practice with the function concept.

NOTE ▶ Solutions to all problems can be found in Chapter 4.

Problem 56
Change the function `fnExample3` so that access to the result of the squaring of the vector ..... is possible.

Problem 57
Change the function `fnExample2` so that the colors for the graphical plots can be given through the function and the signal can then also be plotted accordingly.

Problem 58
Change the function `fnExample2` so that the parameters of the function can be passed on in the form of a parameter structure `fctparams`. The structure fields should then contain the original parameters of the function `fnExample2`.

Problem 59
Write down a MATLAB function that plots a circular disk with a specified radius and returns the circumference and area of the disk as a result. Use the command `axis equal` in order to make the axis partitions equal in length; otherwise, the result is an egg.

---

[15] *Global* parameters can also be defined in MATLAB, but we shall discuss them in detail later in Chapter 2.
[16] There it is realized through pointer variables or references.

The radius and filling color of the disk must be passed on as variables. For simplicity, the center of the circle should always be set at the origin.

When plotting use the MATLAB function `fill` for filling polygonal figures, while you approximate the circle to be drawn with a suitable polygon. Think about how the points on a circle can be expressed mathematically and choose suitable points on the circle so that the polygon resulting from joining the points comes as close as possible to a circle.

Consult the help menu regarding the correct use of `fill`.

1.6.3 MATLAB Language Constructs

As pointed out above, MATLAB is also a programming language and thus is endowed with program language-like constructs. If you type in `help lang` (language) in the MATLAB command window, you will get a survey of these constructs. Some of the results of this command are listed here. The commands that we will not discuss in this introduction are omitted from this display.

```
    Programming language constructs.


    Control flow.
      if          - Conditionally execute statements.
      else        - IF statement condition.
      elseif      - IF statement condition.
      end         - Terminate scope of FOR, WHILE, SWITCH, TRY
                    and IF statements.
      for         - Repeat statements a specific number of times.
      while       - Repeat statements an indefinite number of times.
      break       - Terminate execution of WHILE or FOR loop.
      continue    - Pass control to the next iteration
                    of FOR or WHILE loop.
      switch      - Switch among several cases based on expression.
      case        - SWITCH statement case.
      otherwise   - Default SWITCH statement case.
      try         - Begin TRY block.
      catch       - Begin CATCH block.
      return      - Return to invoking function.

    Evaluation and execution.
      ...
      eval        - Execute string with MATLAB expression.
      feval       - Execute function specified by string.
      ...
```

```
Scripts, functions, and variables.
   ...

Argument handling.
   ...
   nargin      - Number of function input arguments.
   nargout     - Number of function output arguments.
   varargin    - Variable length input argument list.
   varargout   - Variable length output argument list.
   ...


Message display.
   error       - Display error message and abort function.
   ...


Interactive input.
   ...
   pause       - Wait for user response.
   ...
```

MATLAB thus distinguishes six classes of language constructions. For this introduction, the control flow constructs are of particular interest. They are typical of all higher program languages and can be used to control the running of a sequence of commands within a program.

The next example should clarify how these constructs are manipulated. The reader is urged to first type in **help <keyword>** in order to become informed about their syntax.

if Construct – Converting Complex Numbers

A simple example of a conditional distinction that can be handled using the **if** construct is the conversion of complex numbers into their polar (algebraic exponential) representation.

As is generally known, in this conversion a correct calculation of the argument (phase, amplitude) requires a distinction as to which quadrant[17] the complex vector representing the number lies.

---

[17]This is handled in different ways in the literature. In the program **convertComplex** the argument of the complex number is always taken to be a *positive angle* in the counterclockwise sense as measured from the positive real axis. Since the MATLAB function **atan** is defined over the angle between $-\pi/2$ and $\pi/2$, an angle correction must be made in accordance with the quadrant in which the vector lies.

The following MATLAB program **convertComplex** carries out this task and provides the modulus (magnitude) and argument of a complex number, which is entered as a parameter. The argument is given both in radians and in degrees.

```
function [magn, radians, degrees]= convertComplex(cmplxNum)
%
% function convertComplex
%
% call: [magn, radians, degrees]= convertComplex(cmplxNum)
%
% An example of handling the MATLAB if-construct
%
% This example calculates the modulus and argument
% of the complex number cmplxNum specified in the
% algebraic representation (standard for MATLAB)
%
% Input parameter: cmplxNum    Number (real or complex)
%
% Output parameters:   magn       Magnitude (modulus) of cmplxnum
%                      radians    Angle of cmplxNum in radians
%                      degrees    Angle of cmplxnum in degrees

% Calculate the modulus
magn = abs(cmplxNum);

% Calculate the real and imaginary parts

cmplxNumRe = real(cmplxNum);
cmplxNumIm = imag(cmplxNum);

% Determine the quadrant

if cmplxNumRe > 0
   if cmplxNumIm >= 0        % first quadrant
      radians = atan(cmplxNumIm/cmplxNumRe);
   else                      % fourth quadrant
      radians = atan(cmplxNumIm/cmplxNumRe) + 2*pi;
   end;
elseif cmplxNumRe < 0        % second and third quadrants
      radians = atan(cmplxNumIm/cmplxNumRe) + pi;
else                         % special case: real part = 0
   if cmplxNumIm >= 0        % imaginary part positive
```

```
        radians = pi/2;
    else                        % imaginary part negative
        radians = 3*pi/2;
    end;
end;


% recalculating the argument in deg

degrees = radians*180/pi;
```

This program contains many nested **if** constructs so it is somewhat tortuous. Of the various alternatives, the **switch** construct, which we shall discuss below, is better.

This program illustrates the handling of **if** constructs quite extensively, along with an example of **elseif**. The latter makes it possible to distinguish the quadrant of the vector, *including* the special case in which the vector lies on the imaginary axis.

for-loops – Calculating Averages

A simple example of loop programming with a **for** construct is the calculation of the average of a sequence of numbers. The numbers to be averaged are, in typical MATLAB fashion, naturally organized (or to be organized) as a vector. Obviously, MATLAB has a prepared function for calculating averages, the function **mean**, but the following self-written function **averageValue**, should solve this problem with the aid of **for**-loop constructs.

A **for**-loop is always employed when a certain *previously specified* number of operations must be repeated. In this example, the number of the components of the vector (the number of numbers to be averaged) is known.

The following function sums these numbers one after another and then divides the sum by the number of numbers.

```
function [avgval] = averageValue(nums)
%
% function averageValue
%
% call: [avgval] = averageValue(nums)
%
% An example of handling for-constructs in MATLAB
%
% This example calculates the average value
% of the components of the vector Nums.
%
```

```
% input parameter:  nums     The numbers to be averaged,
%                            organized as a vector
%
% output parameter  avgval   the calculated average
%

% calculate the magnitude of the vector
N = length(nums);

% the variable avgval, the average, is
% initialized to 0
avgval = 0;

%  adding the components to avgval, one after the other
for k=1:N             % do this from the 1st to the Nth component
   avgval = avgval + nums(k);
end

% Divide avgval by the number of numbers (N)
avgval= avgval/N;
```

As can be seen in this example, the word **for** must be followed by the condition under which the instructions are to be repeated until the end of the loop (indicated by the word **end**). This condition must cover a finite number of cases. In most cases, as in the example, this will be the upper and lower limits of an integral variable. The loop operation can, moreover, take place in larger steps. Thus, for example, the instruction

```
for k=1:2:N
```

in the above example would cause the calculation to be done only on the odd components.

In conclusion, it should be noted that the above program for MATLAB is not at all typical. Many loops, including this one, can be performed more easily and more elegantly with vectorial constructions in MATLAB (see Problem 61). Of course, there are other situations in which a classical loop construct cannot be avoided.

for-loops – Maximum Value Search

Another simple example of the use of a **for** -loop is the determination of the maximum of a sequence of numbers (for which, of course, there is again a

MATLAB command, namely the command `max`. A similar command (`min`) exists for the minimum.).

Here, we have the number of cases to work through already determined. This makes it a typical task for a `for`-loop.

The following function solves this problem:

```
function [mval] = maxValue(nums)
%
% function maxValue
%
% call: [mval] = maxValue(nums)
%
% An example of handling for-constructs in MATLAB
%
% This example calculates the maximum value
% of the components of the vector Nums.
%
%
% input parameter:    Nums      The numbers to be searched,
%                               organized as a vector
% output parameter:  mval       the calculated maximum

% calculate the magnitude of the vector
N = length(nums);

% the variable mval, the maximum value, is
% initialized to the first value of the variable
mval = nums(1);

% search the components, one after the other
for k=2:N                % do this from 2nd to the Nth component
   if nums(k) > mval     % Compare the current component
      mval = nums(k);    % with the previous maximum: if it is
   end                   % greater, assign this number to mval
end
```

`while`-loop – Input Function

In the preceding examples the number of operations to be repeated was known in advance, so that the running condition could be executed in a `for`-loop. In the above examples this was done concretely by specifying a numerical vector (e.g., `1:2:N`).

In many cases, however, the number of the operations to be carried through identically is not known beforehand. A simple example of this is the reading in of data until a datum satisfies a particular end criterion.

The following simple function **FInput** prompts the user to input data and saves these data in an output vector until an input number is *negative*.

The program uses the existing MATLAB function **input** as an input prompt, together with a so-called **while**-loop for reading in the data.

```
function [InVector] = FInput()
%
% Function FInput
%
% call:  [InVector] = FInput
%
% An example of handling the while-construct in MATLAB
%
% This example prompts the user to
% input data and saves these data in a
% vector [InVector]. This procedure is repeated
% until a NEGATIVE number is read in.
%
%
% input parameter:    none
%
% output parameter:   InVector  values read in

% Prompting the user to input data
% dat contains the input datum
dat = input('Input a number! (End if negative):');
InVector = dat;      % Initialize InVector with this

% Checking and reading in further data, until
% the end criterion is satisfied

while dat >= 0       % as long as the datum is positive:
                     % read in the next number
   dat = input('Input a number! (End if negative):');
                     % append to InVector
   InVector= [InVector; dat];
end;
```

With the **while**-loop employed here, this program will run up to the **end**-instruction belonging to the construction as long as the condition specified after `while` is *true*. In this case the condition is `dat>=0`. Before the first run the condition must be evaluable, which in this case means that at least *one* value must be read in and evaluated. In other programming languages a `do-while` construction exists for these cases; in it the loops will run at least once *in every case*. In MATLAB this kind of behavior has to be imitated by specially evaluating the loop instructions in advance.

By the way, the above example can very easily mislead the beginner into only writing functions for which the parameters are not input from the parameter list but are read in interactively via the `input`-function to the function. Use of this function is, however, rather untypical and for the most part is just for test purposes; it creates a sense that, in general, a function of this sort naturally would not be used within other programs. Thus, you would be better off to put your energy into understanding the parameter-input mechanism of MATLAB.

`switch`-construct – Characteristics of a Vector

When dealing with many cases which are to be distinguished, an `if ...elseif` construction is usually very obscure. In these situations it is better to use a `switch ... case` construction.

The following program uses this type of construction in order to calculate various characteristics of a vector depending on an input parameter (here a string):

```
function [result] = multiFnVector(vector, operation)
%
% function multiFnVector
%
% call: [result] = multiFnVector(vector, operation)
%
% sample call: result = multiFnVector([-1 3 14 2 -3], 'l2norm')
%
% An example of handling the switch-case construct in MATLAB
%
% This example calculates different characteristics
% of a vector, such as its norm or length, etc, depending
% on the string input under the parameter "operation".
%
% input parameters:    vector      real vector
%                      operation   String indicating the
%                                  operation to be performed:
%                                  'l2norm', 'max', 'min',
```

```
%                              'l1norm', 'average',
%                              'maxnorm', 'length'
%
% output parameter    result      calculated value

% Calculating the desired quantity corresponding to operation

switch operation

    case 'l2norm'              % calculate the euclidean norm
        dim = size(vector); % determine the dimension of the
        if dim(2)<dim(1)     % vector when there is no row
          vector = vector'; % vector, then create one
        end;
        result= sqrt(vector*vector');

    case 'max'        % maximum of the values of the components
        result = max(vector);

    case 'min'        % minimum of the values of the components
        result = min(vector);

    case 'l1norm'     % sum of the contributions of the components
        result = sum(abs(vector));

    case 'maxnorm'    % largest magnitude of the components
        result = max(abs(vector));

    case 'average'    % average value of the components
        result = mean(vector);

    case 'length'     % length of the vector
        result = length(vector);

    otherwise
        disp('Please enter "help multiFnVector"');
        error('The input parameter is not allowed!');
    end;
```

Admittedly, this function is not particularly meaningful, since in the above case independent functions would surely be more appropriate, but here we are more concerned with demonstrating the **switch....case** construct than with the problem as such.

It can be seen that, after **switch**, a condition must be entered according to which the cases are to be distinguished. As in this example, it can be a string, but it can also be a number or, in general, an expression. Next, instructions are given for what is to be done in the different cases. These are distinguished by evaluating the expression after the keyword **case**. If the expression shows up in the cases specified under **switch**, then the instructions following it will be executed, otherwise the instruction following **otherwise** will be executed, which in the present case consists of an appropriate error message with the aid of the **error** function. Unlike in C++, no **break** is necessary after each **case** instruction.

In another example we shall use the switch construct in order to write a function which, depending on the *number* of in- and output parameters, will initiate various responses. By the way, it also illustrates the use of the argument functions **nargin** and **nargout**. To do this we modify the earlier program **fnExample.m** from Section 1.6.2:

```matlab
function [t, sinfct, cosfct] = FSwitchIn(f1, f2, damp)
% function FSwitchIn
%
% call: [t, sinfct, cosfct] = FSwitchIn(f1, f2)
% or    [t, sinfct, cosfct] = FSwitchIn(f1, f2, damp)
%
% An example of an MATLAB function with a variable
% number of input parameters

t=(0:0.01:2);

switch nargin
    case 2
        sinfct = sin(2*pi*f1*t);
        cosfct = 2*cos(2*pi*f2*t);
        plot(t,[sinfct; cosfct])
        xlabel('time / s')
        ylabel('Amplitude')
        title('sine and cosine oscillations')
    case 3
        sinfct = sin(2*pi*f1*t);
        cosfct = 2*cos(2*pi*f2*t);
        expfct = exp(-damp*t);
        plot(t,[sinfct; cosfct; expfct])
        xlabel('time / s')
        ylabel('Amplitude')
        title('three gorgeous signals')
```

```
        otherwise
            msg = 'The function FSwitchIn must have 2';
            msg = strcat(msg, ' or 3 input parameters!'):
            error(msg);
    end

    if nargout < 3
        msg = 'The function FSwitchIn should return a time';
        msg = strcat(msg, 'vector and two sine signals!');
        error(msg);
    end
```

With the aid of **nargin** and **nargout** the function can determine the number of parameters used in the call. In this example it has been used to collect certain input combinations and, if necessary, set error messages.

With special cell arrays, **varargin** and **varargout** can, moreover, be used to process input parameter lists and output parameter lists of *variable length*. Here they should always be specified as the *last arguments* in the *definition* of functions. When the functions are *called*, parameter lists of arbitrary length can appear in this position. The interested reader is referred to Problem 67 for a further discussion of this topic.

PROBLEMS

Work through the following problems for practice with MATLAB programming constructs.

NOTE ➤ Solutions to all problems can be found in Chapter 4.

Problem 60
Write a function using a loop construction, which selects the *positive* values in a vector input to the function and returns these values assembled into an output vector (see Section 1.2.3).

Problem 61
Write a function that calculates the standard scalar product of two vectors (see Problem 9) using a **for**-loop.

This solution is not very elegant and is not typical of MATLAB. How would a MATLAB programmer solve this problem?

Problem 62
Write the function **FInput** so that the negative number required to end the input is *not* read into the output vector.

Problem 63
Write a function that plots sin (x) between 0 and 2π in a specified color. The color should be read as a parameter in the form `'red'`, `'blue'`, `'green'`, and `'magenta'`. The different cases for the color are to be distinguished using a `switch...case` construct.

Problem 64
Using a `while`-loop, program the approximation for $\sqrt{2}$ based on Newton's method with the aid of the recursion relation

$$x_{n+1} = \frac{x_n^2 + 2}{2 \cdot x_n}, \qquad x_0 = 2 . \tag{1.3}$$

Repeat until x_n changes only in the 4th decimal place.

Problem 65
With a `while`-loop, write a MATLAB function that provides the quotient of two whole numbers a and b along with the remainder.

For $a = 10$ and $b = 3$, for example, this function should yield the result $q = 3$ with remainder $r = 1$. Thus, it should take place as follows: within the while-loop b will be taken from a as long as possible without the remainder being smaller than b. In the example given here this can obviously be done three times and the remainder is 1.

There is *no need* to verify within the program that a and b are positive whole numbers. When programming, assume that the program will only be called with positive whole numbers.

Problem 66
Write a function that plots two sinusoidal oscillations. The frequencies of the oscillations must be specified as a parameter. Likewise, the plotting range, plot color, and a label for the function must be entered. To do this, the structure `Graphic` in Section 1.3.1 should certainly be used.

Problem 67
Write a function that plots *an arbitrary number* of sinusoidal oscillations (over the same time interval). Thus, the input parameter for the function should include the vector for the (step size) time points, followed by the frequencies of the oscillations. In defining this function use the cell array `varargin` for inputting the variable list of frequencies. With `help varargin` ascertain in what way the contents of the `varargin` lists can be accessed. Reread Section 1.3.2 on cell arrays before doing this.

1.6.4 The Function `eval`

The function **eval** opens up interesting possibilities. With this function MATLAB is able to *evaluate strings* (e.g. commands as MATLAB expressions).

The following example clarifies how **eval** operates:

```
>> clear all
>> whos
>> theCommands = ['x = 2.0; ', 'y = 3.0; ', 'z = x*y; ', 'whos']

theCommands =

x = 2.0; y = 3.0; z = x*y; whos

>> eval(theCommands)
  Name                  Size                    Bytes  Class

  theCommands           1x31                       62  char array
  x                     1x1                         8  double array
  y                     1x1                         8  double array
  z                     1x1                         8  double array

Grand total is 34 elements using 86 bytes

>> z

z =

      6
```

In the variables **theCommands** a sequence of MATLAB commands is defined as a string. This string is interpreted as a MATLAB command sequence via the subsequent **eval** command and obviously correctly executed, as you can see in the result of the multiplication stored as **z**.

In the interactive mode it naturally makes little sense to introduce an **eval** command, since the command can be given anyway. But, within a program, commands can be "assembled" and executed, depending on the program situation.

A brief example illustrates the way it is used.

Data can be *formatted* into a file or written onto the screen using the function **fprintf**. This function is nearly identical to the function of the same name used in C++. The task of the following MATLAB function,

FprintfEval.m, is to print a formatted vector of numbers on the screen with the aid of fprintf. This should yield the format in the form of the number of digits along with the vector.

```
function [] = FprintfEval(x, digits, post)
%
% Function FprintfEval
%
% call: FprintfEval(x, digits, post)
%
% sample call: FprintfEval(x, 10, 7)
%
% Example of a MATLAB function which is used within
% the function eval for constructing commands.
% A vector is to be displayed on the screen as a
% column along with a prescribed format.
%
% input parameters:    x         a vector
%                      digits    width of the number
%                      post      number of places
%                                after decimal point
%
% output parameters:   none

N = length(x);          % set the number of values
digs = num2str(digits); % convert number of places into a string
psts = num2str(post);   % convert number of places after decimal
                        % point into a string

for k=1:N
    printcommand = ...
    ['fprintf(''%', digs,'.', psts, 'f\n'',', num2str(x(k)),')'];
    eval(printcommand);
end
```

The call

```
>> x = [1.2, 3.09, 2.6];
>> FprintfEval(x, 10, 6)
  1.200000
  3.090000
  2.600000
```

shows that the vector $\bar{x}$ is obviously displayed correctly and, as desired, with six places after the decimal point and a width of ten places.

The construction of the print command inside the function core certainly merits comment. First, the syntax of the command **fprintf** requires that the formats be given *as a string*. A normal call in the above example for the k-th component of $\bar{x}$ would look like

```
fprintf('%10.6f\n',x(k))
```

Since the format values have, of course, to be specified numerically, these must first be converted into strings using the function **num2str**. With these strings, in the above solution the complete print command is next assembled *as a string*. The assembly has them written one behind another in a vector (**printcommand**) of characters (data type **char**). The apostrophes in the format string must thus be doubled. Once the string is assembled, it corresponds to the desired command and can be executed using **eval**.

PROBLEMS

Work through the following problems to practice handling **eval**.

NOTE ▸ Solutions to all problems can be found in Chapter 4.

Problem 68
Write a program that generates n random signals of length 10 using the random generator function **rand**, stores these in variables, and then writes these variables individually in ASCII format in text files. The number n of signals should be specified as a parameter in the function. The name of the files should be automatically generated using the number n.

As for using **rand**, consult MATLAB help.

Problem 69
With a bit more thought, the function **FprintfEval.m** can also be written without using the function **eval**. Figure out how.

1.6.5 Function Handles

Under certain circumstances it is appropriate or even mandatory that functions be passed on to other functions *via the parameter lists*. In the following we consider a few examples of this. This technique is of special interest for one of the most important applications of MATLAB (and Simulink), the numerical solution of differential equations, which we shall discuss in Section 1.6.6.

In MATLAB, functions can be passed on to other MATLAB functions across lists of parameters as so-called *function handles*. A *function handle* corresponds to a *pointer* on the function and results when an @ is placed in front of the function involved. Function handles represent a particular data type in MATLAB, as the following sequence of instructions shows:

```
>> clear
>> FH_Sin = @sin;
>> whos
  Name          Size                    Bytes  Class

  FH_Sin        1x1                        16  function_handle array

Grand total is 1 element using 16 bytes
```

The treatment of *function handles* has undergone significant changes in MATLAB 7 compared to earlier versions. Function handles can now be used in the same way as the function names themselves. Thus, in the above example, instead of

```
>> value = sin(2)

value =

    0.9093
```

we can write equivalently

```
>> value = FH_Sin(2)

value =

    0.9093
```

The handle **FH_Sin** is a pointer indicating the function code for *sin*. Thus, this function will be executed when **FH_Sin** is called. It naturally leads to the same result.

In the versions prior to MATLAB 7 the call had to be carried out with the help of the function **feval** in the following form:

```
>> value = feval(FH_Sin, 2)

value =

    0.9093
```

The function **feval** and this call convention are still supported in MATLAB 7. Function **feval** should, however, no longer be used in self-developed MATLAB programs in order to avoid later compatibility problems. Thus, we shall not discuss this function further.

Instead, we shall proceed to the uses to be made of the transfer of a function handle. This can be clarified by an example. To do this we first examine the following source text for the function **fnExample4**:

```
function [integral] = fnExample4(a, b, F, N)
%
% Function fnExample4
%
% call:  [int] = fnExample4(a, b, F, N)
%
% sample call: integ = fnExample4(2, 4, @tstfnct, 3)
%
% An example of manipulating function handles
%
% The present example calculates the integral of
% the function F, whose name is passed on as
% a function handle to fnExample4, over the limits [a,b].
% Simpson's rule is used to calculate the integral;
% for that the range of integration is partitioned
% into two times N intervals.

% partitioning the interval [a,b] (setting the points)

h=(b-a)/(2*N);           % subinterval length
intval=(a:h:a+2*N*h);    % points marking subintervals

integral = F(a)+F(b);    % F at the limits of the interval

                         % F at the subinterval points with
                         % odd indices
for i=3:2:2*N
    integral = integral+2*F(intval(i));
end;
                         % F at the subinterval points with
                         % even indices
for i=2:2:2*N
    integral = integral+4*F(intval(i));
end;

integral = integral*h/3;  % normalizing with h/3
```

As the initial comments show, in this program we are implementing *Simpson's rule* for numerical integration of a real function $F(x)$ over an interval $[a, b]$ with $2n + 1$ partition points ($2n$ partial intervals of length h).

Here, the limits of integration a and b and the function *name* are passed to the function **fnExample4** as *handles*. In order for the Simpson integration to be performed correctly, MATLAB has to find, in its search paths, an m-file (of the same name) in which this function is defined.

As examples, in the accompanying software two functions are defined under the names **tstfnct** and **tstfnct2**.

The corresponding call for **tstfnct2**, which defines the cosine function, accordingly yields seven partition points in the interval $[2, 4]$ for three double intervals:

```
>> simpint = fnExample4(2, 4, @tstfnct2, 3)

simpint =

   -1.6662
```

The exact value is -1.66609992.

Integration by Simpson's rule is an example in which the use of function handles is unavoidable. The program **fnExample4** thus represents the *method* of Simpson integration. It should naturally be applicable to all possible integrable functions. Hence, the function to be integrated may not appear explicitly in the program code. Rather, it has to be passed on to the function in some form similar to a variable. The subinterval points for this are the parameter list and the variables of the function handle.

PROBLEMS

Work through the following problems for practice with manipulating function handles.

NOTE   Solutions to all problems can be found in Chapter 4.

Problem 70
Write a function with which you can calculate numerically the integral of a given real function with two limits a and b using the trapezoid rule.

Modify **fnExample4** to do this.

Problem 71

Write a function with which you can calculate numerically the integral of a given real function with two limits a and b using Simpson's rule *and* the trapezoid rule, where, on one hand, you rely on a comparison of the functions `fnExample4.m` and the solution from Problem 70 and, on the other, on the MATLAB functions `quad` and `trapz`. This function should have no return value, but all the results should be formatted with `fprintf` and displayed on the screen.

1.6.6 Solution of Differential Equations

The (*numerical*) solution of differential equations is one of the great strengths of MATLAB (and Simulink). The techniques for numerical solution of differential equations are considerably refined compared to earlier MATLAB versions. In the same way, the range of methods has been extended. At this point, we can only deal with the more or less direct methods based on the Runga-Kutta procedure for solving initial value problems. Beyond that, however, methods are available for solving boundary value problems, differential equations with delays, and certain partial differential equations. These, however, require extensive mathematical knowledge so that discussing them is beyond the scope of this book.

In the following we shall restrict ourselves to a discussion of the basic solution functions ("solvers") `ode23ode23` and `ode45ode45` based on the Runge-Kutta method and their use for solving ordinary differential equations and systems of ordinary differential equations.

These methods are also fundamental for calculations in Simulink, as we shall see in Chapter 2.

We cannot (and will not) go into the mathematical background of these methods at this point. For this we refer to the relevant mathematical literature.

In order to show how, for example, `ode23` is used for the numerical solution of differential equations, here is an excerpt from the description of `ode23`:

```
>> help ode23

  ODE23  Solve non-stiff differential equations, low order
     method. [T,Y] = ODE23(ODEFUN,TSPAN,Y0) with TSPAN =
     [TO TFINAL] integrates the system of differential equations
     y' = f(t,y) from time TO to TFINAL with initial conditions Y0.
     Function ODEFUN(T,Y) must return a column vector corresponding
```

```
to f(t,y). Each row in the solution array Y corresponds to
a time returned in the column vector T. To obtain solutions
at specific times T0,T1,...,TFINAL (all increasing or all
decreasing), use TSPAN = [T0 T1 ... TFINAL].

[T,Y] = ODE23('F',TSPAN,Y0,OPTIONS) solves as above with
default integration parameters replaced by values in OPTIONS,
an argument created with the ODESET function.  See ODESET for
details.

...
```

The rest of this description deals with other options within this command. These, however, are not relevant at the beginning.

The returned parameters for this function are the calculated time vector T and the corresponding value matrix of the solution(s) Y.

As you can gather from this description, besides the initial conditions $\vec{y}_0$ and the interval ($[t_0, t_{\text{final}}]$) over which the solution is to be calculated, the method **ode23** must be supplied with *one* or *a system* of ordinary differential equations in an m-file. The file name will, in turn, be passed on to the **ode23** method as a *function handle*.

The procedure is similar for **ode45** and the other methods.

We now illustrate this procedure with a (classical) example.

Mathematical Pendulum

We are interested in the solution of the differential equation for the so-called mathematical pendulum,

$$\ddot{\alpha}(t) = -\frac{g}{l} \cdot \sin\left(\alpha(t)\right), \qquad g = 9.81 \, \frac{\text{m}}{\text{s}^2} \,. \tag{1.4}$$

Here, $\alpha(t)$ is the angle that the pendulum (of length l) forms at time t relative to its equilibrium position (see Fig. 1.25).

Since we are dealing with a second order differential equation, its solution requires *two* initial conditions,

$$\vec{\alpha}(0) = \begin{pmatrix} \alpha(0) \\ \dot{\alpha}(0) \end{pmatrix}, \tag{1.5}$$

which define the starting position (deflection) of the pendulum and its initial velocity. These initial conditions are passed on to the integration method as the vector for $\vec{y}_0$.

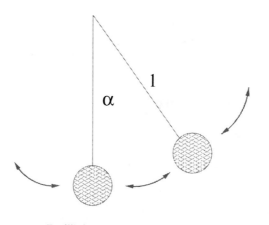

Equilibrium position

FIGURE 1.25   **The mathematical pendulum.**

In order to describe the differential equation, in *MATLAB* we have to define how the derivatives of the unknown function $\alpha(t)$ depend on one another. This, of course, has to be done with the aid of the *numerical values* of the function (i.e., with the vectors of the functions and their derivatives at the subdivision points).

This we do using a MATLAB function **pendde** with which we describe the first and second derivatives and return them as a vector.

The structure of this function must satisfy a definite form, which we shall discuss further in the following. A complete overview can be obtained by entering **odefile** in the search form for MATLAB help.

The function has the following form:

```
function [alphadot] = pendde(t,alpha)
%
% Function pendde
%
% call:  [alphadot] = pendde(t,alpha)
%
% Example of the solution of differential equations
% in MATLAB with ode23
%
% This function defines the differential equation
% for a mathematical pendulum (l=pendulum length).
% For MATLAB to proceed with this, the second order DE
% must initially be converted into a system of first order DEs.
```

```
%% Setting constants

l=10;                          % pendulum length
g=9.81;                        % acceleration of gravity m/s

%% Preliminary initialization

alphadot = [0;0];

%% Representation of the differential equation

                               % the first first order equation
alphadot(1) = alpha(2);

                               % the second first order equation
alphadot(2) = -(g/l)*sin(alpha(1));
```

This implementation obviously requires some comment.

First, we should note the following: in order for MATLAB's **ode23** to handle the Eq. (1.4), the equation has to be converted into a *system of first order differential equations!*

To do this, we set

$$\alpha_1(t) := \alpha(t), \tag{1.6}$$

$$\alpha_2(t) := \dot{\alpha}(t). \tag{1.7}$$

Eq. (1.4) can then be rewritten in the form

$$\dot{\alpha}_1(t) = \alpha_2(t), \tag{1.8}$$

$$\dot{\alpha}_2(t) = -\frac{g}{l} \cdot \sin(\alpha_1(t)) \tag{1.9}$$

with the initial conditions

$$\vec{\alpha}(0) = \begin{pmatrix} \alpha(0) \\ \dot{\alpha}(0) \end{pmatrix} = \begin{pmatrix} \alpha_1(0) \\ \alpha_2(0) \end{pmatrix}. \tag{1.10}$$

It is *this representation* to which **ode23** and the accompanying representation of the differential equation connect in the m-file **pendde.m**.

The equation **alphadot(1) = alpha(2);** thus represents the first part of the system of equations (1.8) and **alphadot(2) = -(g/l)*sin(alpha(1));** the second; this means that at each time point, the values on the right-hand side are passed on to the components of the vectors representing the derivatives.

The parameter **t** *must* also be passed on to **pendde**, even if it is not used within the function. It is required by **ode23**. Function **ode23** processes the initial conditions first.

All differential equations to be solved using **ode23** (or other methods) essentially have to be brought into the form *"vector of the first order derivatives equals a function of the solution vector"* and so represented in a MATLAB file.

The following calls then provide a representation of the solution for the initial values $\alpha(0) = \frac{\pi}{4}$ and $\dot{\alpha}(0) = 0$ within the interval $[0, 20]$ and for a pendulum of length 10. The solution and its derivative (dashed curve) are shown in Fig. 1.26.

```
>> [t, solution] = ode23(@pendde, [0, 20], [pi/4,0]);
>> plot(t, solution(:,1),'r-',t, solution(:,2),'g--')
```

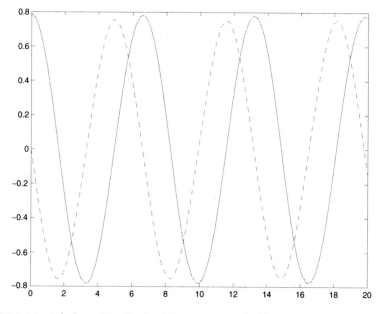

FIGURE 1.26 **Solution of Eq. (1.4) with MATLAB's ode23.**

The following comments pertain to the calculated solution:

The solution algorithm works within a so-called *step-size control*. This means that where the solution varies little, *large* step sizes are used, and where it varies rapidly, *small* step sizes are used. This increases the efficiency of the algorithm. A look at the work space for the above example shows how the step size (the vector **t**) was consecutively adjusted:

```
>> [t, solution]

ans =
```

0	0.7854	0
0.0001	0.7854	-0.0001
0.0007	0.7854	-0.0005
0.0036	0.7854	-0.0025
0.0180	0.7853	-0.0125
0.0901	0.7826	-0.0624
0.2372	0.7659	-0.1634
0.4390	0.7193	-0.2976
0.6822	0.6283	-0.4470
0.9558	0.4857	-0.5894

```
...
```

Still, it is often worthwhile to derive a solution with an *equidistant* step size. In that case the vector on which the solution is to be evaluated can be explicitly given in the call for **ode23** :

```
>> [t, solution] = ode23(@pendde, (0:0.1:20), [pi/4,0]);
>> [t, solution]

ans =
```

0	0.7854	0
0.1000	0.7819	-0.0693
0.2000	0.7715	-0.1381
0.3000	0.7543	-0.2059
0.4000	0.7304	-0.2723
0.5000	0.6999	-0.3366
0.6000	0.6631	-0.3985
0.7000	0.6202	-0.4572
0.8000	0.5717	-0.5123
0.9000	0.5178	-0.5631
1.0000	0.4592	-0.6091
1.1000	0.3961	-0.6498
1.2000	0.3293	-0.6845
1.3000	0.2593	-0.7130
1.4000	0.1868	-0.7346

The step size (see the vector **t**) is now, as desired, uniform with a separation of 0.1.

Another important comment touches on the fact that in this example the parameters g and l required in the description function **pendde** are defined with fixed values.

Thus, in the present version g and l can be varied only if the source text of **pendde** is modified. This cannot, of course, be the answer if, for example, the pendulum length is to be varied in the course of a multiple simulation.

Nevertheless, it is possible in principle to provide other parameters to the definition functions for the differential equations. The interested reader can turn to the related MATLAB documentation or the MATLAB help to examine this question and, if necessary, for dealing with Problem 77.

RC Low-pass Filter

We conclude this section with another, somewhat simpler example.

Fig. 1.27 is a circuit diagram of an RC combination with a voltage source.

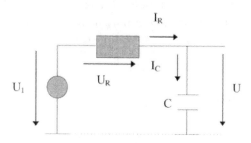

FIGURE 1.27   **RC combination with a voltage source.**

The voltage at the output of the system obeys the following first order, linear differential equation:

$$\frac{d}{dt}u(t) = -\frac{1}{RC}u(t) + \frac{1}{RC}u_1(t). \qquad (1.11)$$

We solve this differential equation numerically using MATLAB's **ode23** function, first by representing the differential equation in a MATLAB function named **RCcomb** as follows:

```
function [udot]= RCcomb(t,u)
%
% Function RCcomb
%
% call:  [udot] = RCcomb(t,u)
%
% Example of the solution of differential equations
```

```
% in MATLAB with ode23
%
% This function defines the differential equation
% for an RC combination with voltage source u1(t).

R = 10000;                    % resistance R
C = 4.7*10e-6;                % capacitance C

udot = 0;                     % preinitialization

                              % the equation is of first order
                              % u1(t) must be defined as
                              % a MATLAB function

udot = -(1/(R*C))*u + (1/(R*C))*u1(t);
```

The function $u_1(t)$ must be defined as a MATLAB function. In this example we have defined the function u_1 as the unit step function. It is executed as follows in the m-file **u1.m** in the working directory:

```
function [step] = u1(t)
%
% function u1
%
% call: [step] = u1(t)
%
% Implementation of the unit step function
%
% ...

step = t>=0;    % This is so simple!
```

The MATLAB commands

```
>> [t, solution] = ode23(@RCcomb, [0, 3], 0);
>> plot(t, solution)
>> grid
```

then provide the solution shown in Fig. 1.28 in the time interval $[0, 3]$ s. Here, fixed values of 10 kΩ and 4.7μF, which yield a time constant of $\tau = RC = 0.47$ s, have been used.

This value can be read very nicely off the graph, since it is known that in that time the response to a step function reaches $(1 - e^{-1})$ times, or 63 % of the height of the step (here 1).

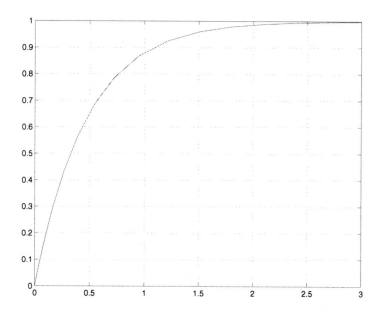

FIGURE 1.28 **Step function response of the RC combination for 10 kΩ and 4.7 μF.**

PROBLEMS

Work out the following problems for practice in solving differential equations with MATLAB.

NOTE ► Solutions to all problems can be found in Chapter 4.

Problem 72

Write a function with which the differential equation of the mathematical pendulum in its *linearized* form,

$$\ddot{\alpha}(t) = -\frac{g}{l} \cdot \alpha(t) \tag{1.12}$$

can be solved.

Modify `pendde.m` accordingly.

Then compare the two numerical solutions for large initial deflections.

Problem 73

Determine the solution of Eq. (1.11) for this RC combination with a voltage source for different excitation functions (source voltages) $u_1(t)$.

Problem 74

Solve one of the examples of differential equations using the procedure `ode45` and compare this solution with that from `ode23`.

Problem 75

Solve the following differential equation for a growth process using MAT-LAB:

$$\dot{P}(t) = \alpha(P(t))^\beta \quad \text{with} \quad \alpha = 2.2,\ \beta = 1.0015. \tag{1.13}$$

Problem 76

Use MATLAB to solve the differential equation for the so-called VZ1-element (first-order delay element),

$$\frac{d}{dt}v(t) + \frac{1}{T}v(t) = \frac{K}{T}u(t). \tag{1.14}$$

At the same time, make some reasonable assumptions about K and T, with which you can propose a probability interpretation.

Problem 77

Solve the differential equation for the mathematical pendulum (Eq. 1.4) for different pendulum lengths using MATLAB. It should be possible to transfer the respective pendulum lengths by calling the solution algorithm (`ode23`).

In addition, after checking with MATLAB help, modify `pendde.m` in a way such that other parameters can be passed on through the definition function for the differential equation.

Problem 78

Solve the system of differential equations

$$\dot{y}_1(t) = -2y_1(t) - y_2(t) \qquad y_1(0) = 1, \tag{1.15}$$
$$\dot{y}_2(t) = 4y_1(t) - y_2(t) \qquad y_2(0) = 1 \tag{1.16}$$

with the aid of MATLAB. Compare the solution with the exact solution, which you can calculate by hand or by using the symbolics toolbox (see Section 1.8).

Problem 79

If *only one* return value is specified, the functions `ode..` yield a *structure* that contains all the necessary statements about the solution. This structure can be processed by the function `deval`.

Familiarize yourself with the syntax of `deval` and solve Eq. (1.4) once again using `ode23`. Next, evaluate the result using `deval` for equidistant partition points separated by 0.01. Plot the solution using the return value of `deval`.

1.7   MATLAB EDITOR AND DEBUGGER

The *editor-debugger* integrated into the MATLAB program package has again been substantially improved in terms of ease and functionality in MATLAB 7. The capabilities of this tool, which we have used throughout Section 1.6, will be reemphasized here. Unfortunately, at this point we can only provide a very brief overview of the capabilities of the editor-debugger. We limit ourselves to a few topics that will probably be of use to the beginner. Advanced users will find that many more functions are available.

1.7.1   Editor Functions

As noted in Section 1.6.1, the editor is started up when an m-file is opened or newly set up using the menu commands **File - open ...** or **File - new**. The integrated Editor-debugger will be automatically employed in MATLAB, unless another editor has been installed in **Preferences**, but there is generally no need for that and it should be avoided.

The editor is also opened up automatically by double clicking on an m-file in the **Current directory** window or by right-clicking **Open** in the context menu.

Likewise, the editor is opened and an m-file created if you mouse click a group of commands in the **History** window or (alternatively) press the **Shift-** or **Control-** key and then right click **Create m-file** in the context menu. This is especially interesting for the simple creation of script files with which successful MATLAB command sequences can be assembled into meaningful units at the end of a MATLAB session.

After opening, the editor shows up in the form shown in Fig. 1.29. (Here, the file **pendde.m** is open as an example.)

Of course, the MATLAB editor has the important features customary in the editors of other high language development systems, such as color highlighting of keywords, comments, indenting, etc. There is no need to discuss this further.

The important elements of the editor interface are:

1. the editor window,
2. cells,
3. menu lists,
4. cell toolbar,
5. icons for window configuration, and
6. icons for debugging, and function stack.

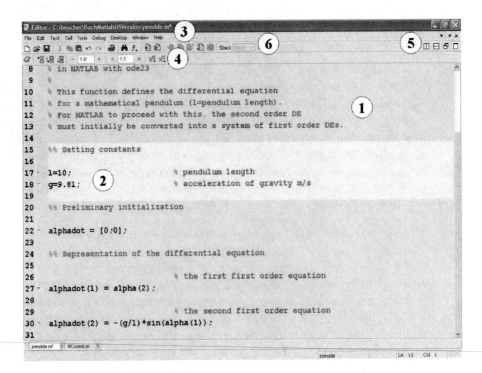

FIGURE 1.29   Editor-debugger with the `pendde.m` file open.

The editor window (1) is useful for the input of the (program) text. Style, font, and color can be set using **Preferences**.

Cells represent an important innovation. Cells are sections that can open with a comment line, which begin with *two* comment symbols (%%) and extend to the next line, also beginning with (%%) or to the end of the file. If the cursor is positioned in a cell, the cell will be highlighted in color (provided the cell-mode is switched on in the **Cell** menu).

Cells are useful in program development. By using the menu command in the **Cell** menu or using the icons in the cell toolbar (4), cells can be specially executed (**Evaluate Cell** icons in front). In this way, individual variables can be changed (either in the Edit fields in (4) or in the workspace), and the operation of the code can be tested. In addition, cells simplify the documentation of m-files (only scripts). Various formats for document-ing an m-file can be chosen via the menu entry **File - Publish to**. The cell definition lines (%%) will be taken over as headings in this documentation.

It should be made clear here that cells are primarily intended for the development of script files. Of course, cells can also be used in MATLAB functions, but in that case all the variables employed must be defined in the workspace during the development phase.

Besides the customary standard entries for windows editors, entries for program development and debugging (**Cell**, **Tools**, **Debug**) can be found in the menu list (3). The icons (5) can be used to divide the screen into several work windows when more than one file is to be processed. The icons (6) for selecting debugging functions will be discussed in the next section.

1.7.2 Debugging Functions

Basically two types of errors show up during program development: *syntax errors* and *runtime* errors. Syntax errors are usually easy to rectify, since the programmer will be advised by an appropriate error message. Runtime errors occur in a syntactically correct program during execution and manifest themselves either as a sudden termination of the program (a crash) or in obviously false results.

In those cases it may be necessary to examine closely the running of the program *during execution*. This task is undertaken by a so-called *debugger*, a program containing so-called *breakpoints* at specific places where the execution of the program being developed can be stopped and inspected.

Breakpoints can be set or eliminated using the menu entry under **Debug** or by clicking on the icon with the red point.

The MATLAB debugger recognizes three different kinds of breakpoints, of which only the so-called *standard breakpoint* will be discussed here.

A (standard) breakpoint is indicated by a red point[18] in the source text. The breakpoint is set in the line where the cursor stands.

When the program is executed this causes an automatic change in the editor and the program is brought to a halt at the position of the breakpoint. The variable values registered up to that point can then easily be viewed, by pointing at the corresponding variables with the mouse (without clicking). The values are displayed in an automatically deployed window.[19]

Fig. 1.30 shows a breakpoint set in **pendde.m** with the mouse pointed at one of the variables, along with an opened variable value window.

Alternatively, the variables can also be displayed in the command window or by clicking on the context menu in the array editor.

---

[18]It should be noted that for this, the file must be saved with complete changes; otherwise, the breakpoint will be indicated in grey.

[19]It is assumed that this functionality is set in **Preferences**.

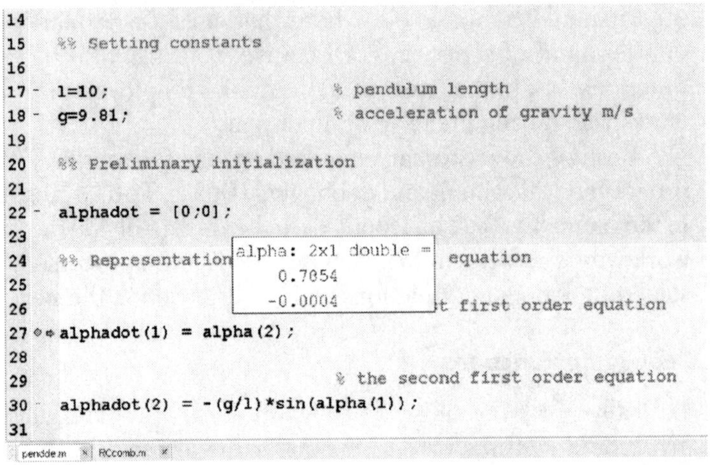

```
14
15    %% Setting constants
16
17 -  l=10;                          % pendulum length
18 -  g=9.81;                        % acceleration of gravity m/s
19
20    %% Preliminary initialization
21
22 -  alphadot = [0;0];
23
24    %% Representation  alpha: 2x1 double =  equation
25                         0.7054
26                        -0.0004            t first order equation
27 ◇* alphadot(1) = alpha(2);
28
29                                   % the second first order equation
30 -  alphadot(2) = -(g/l)*sin(alpha(1));
31
```
pendde.m   RCcomb.m

FIGURE 1.30   Editor-debugger with the `pendde.m` file open, a breakpoint and the displayed variable content.

Of course, the values of the variables are not just displayed. The values can also be *changed* for the runtime of the program, which is very useful for testing.

In the command window the debug mode can be recognized by a modified prompt,

    K>>

It is best to change the variables at this point. Thus, for example, after the stop at the breakpoint shown in Fig. 1.30 the components of the variables `alpha` can be set to 0.5 and 0.1 using the command

    K>> alpha = [0.5;0.1]

If the program is then run in the debug mode (see the menu entry **Debug**), further calculations will use *these* values.

Of course, as we have already explained in Section 1.6.2, each function has *its own memory domain*. Only those variables (see `whos`) that belong to the so-called *function stack* of the function within which the program stops are available. The memory domain can indeed be changed, as needed, in the pull-down menu **Stack** (6).

Readers are urged to become familiar with the capabilities of the debugger. This relatively small effort will be amply repaid if they want to develop larger MATLAB programs.

Besides the debugger, there are other tools for supporting program development: code checks with *M-Lint* and the *profiler*.

Extensive runtime analyses can be carried out with the profiler, but these are clearly useful only to the advanced user. With the M-Lint program (in the menu `Tools - Check Code with M-Lint`) a statement can be set in which the programmer is informed, for example, about unused variables, syntax errors, or other absurdities in the code. This can also be useful for the beginner.

1.8   SYMBOLIC CALCULATIONS WITH THE SYMBOLICS TOOLBOX

The fundamental difference between algebraic and numerical simulation tools has already been pointed out in the introduction. MATLAB is a *numerical* simulation tool and draws its strength from this fact. Of course, it is frequently also useful to be able to perform symbolic calculations. It would be an advantageous for the MATLAB user to be able to do this without leaving MATLAB, since otherwise he would have to learn the unfamiliar command syntax of a computer algebra program, such as MAPLE or MATHEMATICA. The *symbolics toolbox* of MATLAB satisfies this purpose. This toolbox essentially involves an adaptation of the core of MAPLE in MATLAB syntax. At this point we can only discuss the capabilities of this toolbox very briefly, since an extensive description would go beyond the scope of this book. A first overview of these capabilities can be obtained by entering the command `help symbolic`. The following excerpt from the response to this command shows which main command groups are available. Then we shall consider a few important examples.

```
Calculus.
   diff        - Differentiate.
   int         - Integrate.
   limit       - Limit.
   taylor      - Taylor series.
   ...


Linear Algebra.
   ...
   eig         - Eigenvalues and eigenvectors.
   ...
   poly        - Characteristic polynomial.
```

```
Simplification.
  simplify    - Simplify.
  ...
  subs        - Symbolic substitution.

Solution of Equations.
  solve       - Symbolic solution of algebraic equations.
  dsolve      - Symbolic solution of differential equations.
  ...


...
Basic Operations.
  sym         - Create symbolic object.
  syms        - Short-cut for constructing symbolic objects.
  pretty      - Pretty print a symbolic expression.    ...


...


Access to Maple. (Not available with Student Edition.)
  maple       - Access Maple kernel.
  ...
```

In order to be able to execute a symbolic calculation, it is necessary to communicate to MATLAB that the variables to be used in the following commands are *symbols*, and *not*, as is usual in MATLAB, *numerical* variables. With the command

```
>> syms x y v
```

for example, the *symbols* x, y, and v are established. A look at the workspace confirms this:

```
>> whos
  Name        Size          Bytes  Class
  v           1x1             126  sym object
  x           1x1             126  sym object
  y           1x1             126  sym object
Grand total is 6 elements using 378 bytes
```

A symbolic expression in these variables can now be processed using the appropriate commands from the toolbox. The following example differentiates the function

$$f(x, y) = \sin(xy^2) \cos(vxy)$$

with respect to *y* or *v*:

```
>> f = sin(x*y^2)*cos(v*x*y)        % defining the function

f =

sin(x*y^2)*cos(v*x*y)

>> % differentiate with respect to symbol y
>> dfy = diff(f,'y')

dfy =
2*cos(x*y^2)*x*y*cos(v*x*y)-sin(x*y^2)*sin(v*x*y)*v*x

>> % differentiate with respect to symbol v
>> dfv = diff(f,'v')

dfv = -sin(x*y^2)*sin(v*x*y)*x*y
```

The interesting thing in this example is that the MATLAB function `diff` is given twice, namely as an "ordinary" MATLAB function (see `help diff`) and as an "overloaded" symbolic function (see `help sym/diff`). This technique makes it possible to use a natural name assignment, although at a price, in that the user has to pay attention to exactly which of the functions is meant.

The preceding example reveals an important disadvantage of the symbolics toolbox. In most cases, the output of results is very obscure and hard to read. It is highly recommended that the command `pretty` be employed, in order to make the output in the command window readable; for example,

```
>> f = sin(x*y^2)*cos(v*x*y);
>> pretty(f)

                              2
                         sin(x y ) cos(v x y)
>> dfy = diff(f,'y');
>> pretty(dfy)

                 2                        2
        2 cos(x y ) x y cos(v x y) - sin(x y ) sin(v x y) v x

>> dfv = diff(f,'v');
>> pretty(dfv)

                            2
                     -sin(x y ) sin(v x y) x y
```

In essence, as noted above, the symbolics toolbox makes the functionality of MAPLE available to the user. The graphics commands are an exception to this. Anyone who has the full version of MATLAB and the toolbox and also knows their way around MAPLE will find the entire world of MAPLE available through the command **maple**. With this command it is possible to set a MAPLE command in MATLAB in the original syntax. For example, the second derivative with respect to x of the above example function $f(x, y)$ can be determined in the following way:

```
>>                              % translating f into MAPLE syntax
>> maple('f := sin(x*y^2)*cos(v*x*y);')

ans =

f := sin(x*y^2)*cos(v*x*y)

>> df = maple('diff(f,x$2);')   % derivative in MAPLE syntax

df =

-sin(x*y^2)*y^4*cos(v*x*y)-2*cos(x*y^2)*y^3*sin(v*x*y)*v
                        -sin(x*y^2)*cos(v*x*y)*v^2*y^2

>> df = sym(df)         % conversion of string into symbol;
                        % necessary as df is initially a string
                        % (see: whos)

df =

-sin(x*y^2)*y^4*cos(v*x*y)-2*cos(x*y^2)*y^3*sin(v*x*y)*v
                        -sin(x*y^2)*cos(v*x*y)*v^2*y^2

>> pretty(df)           % pretty-print the result

           2    4                         2    3
  -sin(x y ) y  cos(v x y) - 2 cos(x y ) y  sin(v x y) v

                    2               2  2
      - sin(x y ) cos(v x y) v   y
```

This capability is unfortunately not available in the student version of MATLAB.

1.8.1 Symbolic "Auxiliary Calculations"

One very useful application of the symbolics toolbox in many calculations is the calculation of symbolic expressions at intermediate times in purely numerical computations. An example would be the exact calculation of an integral, rather than a numerical approximation, when the functional expression for the integrand is available. The following simple example from electronics should make this application clear. The so-called *effective (RMS) value* of a rectified sinusoidal voltage of amplitude 1 is to be calculated. The effective (root mean square, RMS) value of an alternating current can be interpreted as the DC voltage, which has the same power as the AC current. For a periodic signal $f(t)$ of duration T seconds, it is defined, in general, as

$$U_{eff} = \sqrt{\frac{1}{T}\int_0^T f^2(t)\, dt} \ .$$

Compare this with Fig. 1.31.

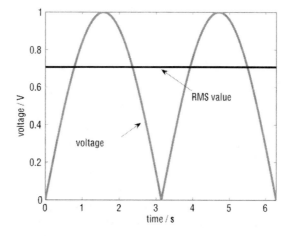

FIGURE 1.31   Rectified sine wave and the RMS equivalent voltage.

This figure shows a rectified sine wave of period $T = 2\pi$ s and the RMS voltage associated with it.

In order to calculate this using MATLAB, we enter the following MATLAB sequence:

```
>> T = 2*pi;              % set the period
>> t=(0:0.5:2*pi);       % subdivide the time interval
```

```
>> f=sqrt(sin(t).^2);        % define the rectified sine
>> IntU = trapz(t,f.^2);     % obtain the integral using the
                             % trapezoid rule for integration
                             % (This is an approximation
                             % of the integral.)
>> Ueff = sqrt((1/T)*IntU);  % calculate the RMS value
```

Here, we obtain the following approximation of the RMS value using a numerical integration technique (the trapezoid rule with the MATLAB function `trapz`):

```
Ueff =

    0.7050
```

The errors associated with this approximation and, thereby, those associated with the discrete partitioning of the time interval, must now be calculated.

In this simple example the RMS value can, in fact, be calculated *exactly*, once the capabilities of the symbolic toolbox are employed. Here, the solution looks like this:

```
>> T = 2*pi;                 % set the period
>> t=(0:0.5:2*pi);           % subdivide the time interval (but
                             % this time only for the plot)
>> f=sqrt(sin(t).^2);        % Define the rectified sine wave
                             % (for the plot)
>> syms x P                  % Define symbolic quantities
                             % (not t and T; otherwise time vector
                             % and period T would be overwritten)
>> F = sqrt(sin(x)^2);       % Define signal symbolically
                             % Integrate the integral SYMBOLICALLY
                             % with respect to x over [0,P]
                             % (This is the EXACT value of
                             % the integral)
>> UeI = int( (1/P)*F^2,x,0,P)

UeI =

1/2*(-cos(P)*sin(P)+P)/P

>> Ue = sqrt(UeI);           % Calculate RMS value exactly
                             % Now the actual value
                             % of T is substituted for P (using
                             % the substitution function subs)
```

```
Ueff = subs(Ue,P,T)      % gives the numerical value

Ueff =

    0.7071
```

Now the exact value of **Ueff** can be calculated numerically. The exact value is, in general, always $\frac{1}{\sqrt{2}}$ for a sinusoidal oscillation of amplitude 1. This is completely independent of the period.

In the above example, you can also see the numerical errors in the approximation with **trapz**, which originate in the excessively coarse subdivision (here 0.5) of the range of integration.

In general, symbolic conversions of this type are often advantageous, especially for calculating derivatives (see Problem 83) or for solving simple differential equations.

PROBLEMS

Work through the following problems involving the symbolic toolbox. Beforehand, make a precise study with **help ...** of the syntax of the symbolic functions that you might use.

NOTE ▶ Solutions to all problems can be found in Chapter 4.

Problem 80
Integrate the function

$$g(x) = \sin(5x - 2) \tag{1.17}$$

twice symbolically (generation of antiderivatives).

Problem 81
Determine the 3rd order Taylor expansion term for the function $g(x)$ about the expansion point $x_0 = 1$.

Problem 82
Solve the differential equation

$$\dot{y} = xy^2 \tag{1.18}$$

symbolically using the command **dsolve**.

Problem 83

Plot the function from Problem 80 along with its first and second derivatives over the interval [0, 10] superimposed in the same plot window in different colors.

Problem 84

For readers with knowledge of MAPLE and the complete version: Integrate the function from Problem 80 twice symbolically (generation of antiderivatives), employing the functionality of MAPLE.

Chapter 2

INTRODUCTION TO SIMULINK

S imulink is a tool for simulating *dynamic systems* with a graphical interface specially developed for this purpose. Within the MATLAB environment, Simulink is a *MATLAB toolbox* that differs from the other toolboxes, both in this special interface and in the special "programming technique" associated with it. There is a further difference, in that the source code of the Simulink system is not open, but this is of no concern for our purposes. The goal of this chapter is to introduce *simple* manipulations with Simulink and to clarify the interaction of Simulink with MATLAB.

2.1   WHAT IS SIMULINK?

As noted above, dynamic systems can be simulated using *Simulink*. In the great majority of cases this means *linear or nonlinear time-dependent processes* that can be described using *differential equations* or (in the case of discrete times) difference equations. Another common way of describing dynamic systems is with *block diagrams*.

This is an attempt to understand the behavior of the system by means of a graphical representation, which essentially consists of representations of individual components of the system together with the signal flow between these components. Fig. 2.1 is an example (for the first-order delay element VZ1; see Section 1.6). Simulink is based on this form of representation. A graphical interface is used to convert a block diagram of this sort (almost) directly into Simulink and simulate the operation of the system. We note that a well-grounded use of Simulink requires some knowledge of control technology and systems theory that is beyond the scope of this introduction. Therefore, here the treatment of Simulink will be restricted to a central theme, specifically the *numerical solution* of simple *differential equations*.

As noted above, dynamic systems (that are continuous in time) will be described by differential equations. Thus, when we describe the system with a block diagram and simulate the reaction of the system to an input signal, we

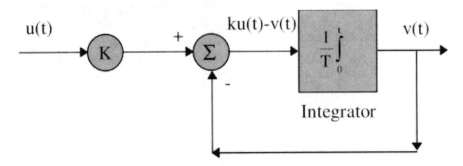

FIGURE 2.1 Block diagram representation of a dynamic system.

are essentially looking for nothing more than the *solution of the differential equation underlying the system.*

It is also possible to convert a differential equation into a block diagram and to solve the differential equation numerically with Simulink. Simulink is, therefore, a *numerical differential equation solver.*

Section 2.1 shows how to use Simulink for this purpose.

2.2 OPERATING PRINCIPLE AND MANAGEMENT OF SIMULINK

The Simulink program is started with the command `simulink` or the command `open_system('simulink.mdl')` in the MATLAB command window. In the first case, a so-called *Block Library Browser* (see Fig. 2.2) is opened. This displays the available blocks in the Simulink library in the form of a list or in the form of icons in the customary mode of Windows Explorer or other modern compilers.

The block library is organized into functional groups, for example for the production of signals (functions), as `Sources`, or blocks of functions for nonlinear operations, as `Nonlinear`.

In the second case, that is on calling `open_system ('simulink. mdl')`, a window appears in which the symbols for the different classes of blocks of functions are only displayed in the form of icons.

Of course, it is a matter of taste which form of opening you choose for Simulink. In the following we prefer the first solution, since it is easier to input the command `simulink`. Once a list entry or icon is selected, a window section opens up in the `Simulink Library Browser` with the function symbols contained in the function library. Clicking on `Sources`, for example, opens up the window shown in the background in Fig. 2.2,

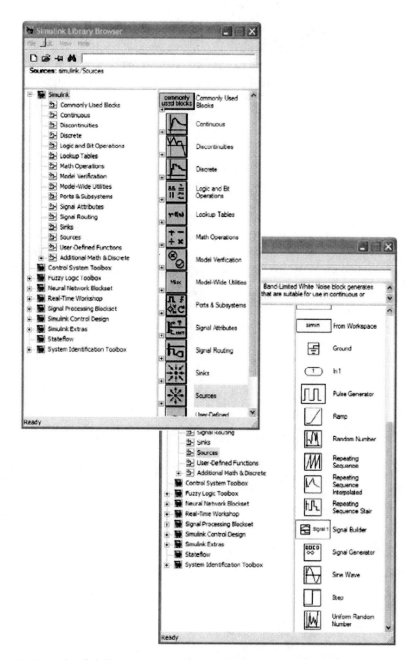

FIGURE 2.2   **Simulink library browser (foreground) and the opened Simulink function library** Sources **(background).**

which displays a selection of the function blocks available to Simulink for generating signals. There, for instance, you can the blocks **Sine Wave** for generating a sine wave or **Pulse Generator** for generating a rectangular pulse. This system can be expanded by the user through adding self-defined function blocks. We discuss this capability briefly in Section 2.4.

2.2.1 Constructing a Simulink Block Diagram

If you then want to construct your own simulation system using the block libraries, your first have to open an empty window by selecting **File - New Model** in the Simulink browser (Fig. 2.2). Already existing block diagrams can be opened up under their file names using **File - Open**. It is recommended that the empty window be saved immediately under a suitable file name as an mdl-file (mdl=model) using **File - Save As**.

In the following example we name the file **s_test1.mdl**. Now, with the mouse we drag a block (e.g., the **Sine Wave** block) out of the block library browser into the empty window. If you don't want to use the name "sine wave" (perhaps because there are many similar blocks in the system and the blocks are to be distinguished by name), then you can click with the mouse on the text line "sine wave" and edit the name from the keyboard. In this way we can rename the block "source signal," for example. The system **s_test1** then has the form shown in Fig. 2.3. The **Sine Wave** block has also been slightly enlarged by dragging at the corner.

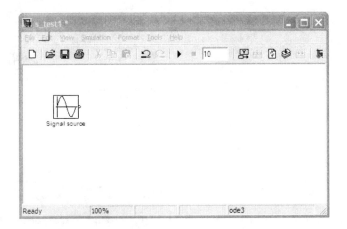

FIGURE 2.3 **The Simulink system** s_test1 **after insertion of the** Sine Wave**-Block.**

In the next step we want to *integrate* the source signal. To do this we open the function library **Continuous**. We drag the **Integrator** block from this library to the system window of **s_test1** and connect the output of the source signal block to the input of the integrator using the mouse. This takes some practice at the beginning. It is always best to draw the connecting line opposite to the signal propagation direction from the input of the target block to the output of the source block; that is, from the integrator to the source signal in this example. The system then has the form displayed in Fig. 2.4.

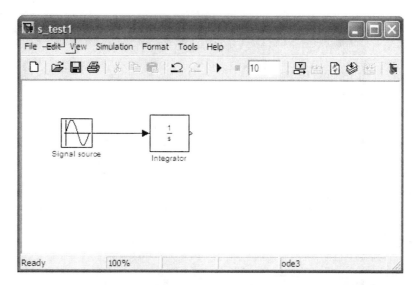

FIGURE 2.4 **The Simulink system s_test1 after insertion of the** Sine Wave**-Block and the** Integrator**-Block.**

The reader should not be bothered by the entry $\frac{1}{s}$ in the integrator block. This relates to the *Laplace transform* of the integration. Most of the linear function blocks are characterized by the Laplace transform or Z-transform (the discrete counterpart of the Laplace transform). Here we will not dwell on these transforms, since in this introduction we basically only need the integrator block.

We now extend the test system **s_test1** so that the sine signal and its integral can be seen in a single window. To do this we first choose a

Multiplexer block[1] **Mux** from the **Signal Routing** library and add it to the system, and choose the block **Scope** from the Sinks library. The source signal and the integrator output are finally connected by mouse movements to two inputs of the multiplexer and the output of the latter, to the input of the **Scope** block. When connecting "around a corner," as with the signal passing from the source signal to the multiplexer, the mouse button has to be released twice along the way. Note also that a connection to a signal path (here between the source and integrator) is only correct if a small rectangular point appears at the crossing point.

The (almost) ready system can be seen in Fig. 2.5.

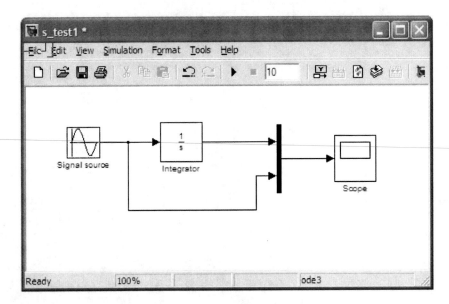

FIGURE 2.5   **The (almost) ready Simulink system** **s_test1.**

Before we can perform a simulation using the block diagram, however, a few work steps have to be taken. On one hand, the *parameters of the blocks to be used* must be set correctly —for example, where does Simulink find out what frequency your sine signal should have? On the other hand, such things as the duration of the simulation, numerical solution procedures, etc., (i.e., the *general parameters of the simulation*) have to be established. And, last but not least, the simulation model should be provided with at least a

---

[1]Its task is to combine multiple signals into a single vector-valued signal.

minimum of comment or labels, since well-organized documentation is also essential in graphical programming.

2.2.2 Parametrizing Simulink Blocks

We begin with block parametrization. As an example, the `Mux` block used in the system `s_test1` obviously has two inputs, which just happens to be right for our purpose. The number of inputs can, in general, be set. As is customary for modifying block parameters, we open up the parameter list belonging to the block by double-clicking on the corresponding block symbol. Double-clicking `Mux` yields the window shown in Fig. 2.6

FIGURE 2.6   **The parameter window for the Simulink block** Mux.

The number of inputs for the multiplexer can be freely specified there in the entry under `Number of inputs`.

The `Display option` parameter is also of interest. Here, the graphical representation of the multiplexer can be changed. The representation set there, `bar`, shows the multiplexer as a thick line, which is especially suitable for many inputs. With only two inputs, as in our case, it may be better to set `signals`, for in this case the input ports of the block, after enlargement of the symbol with the mouse, will be supplied with the names of the signals (See the comments on labeling the signal arrows below), which greatly increases the clarity, especially for large simulation systems.

We parametrize the sine signal in similar fashion. Double-clicking its block opens up the window shown in Fig. 2.7.

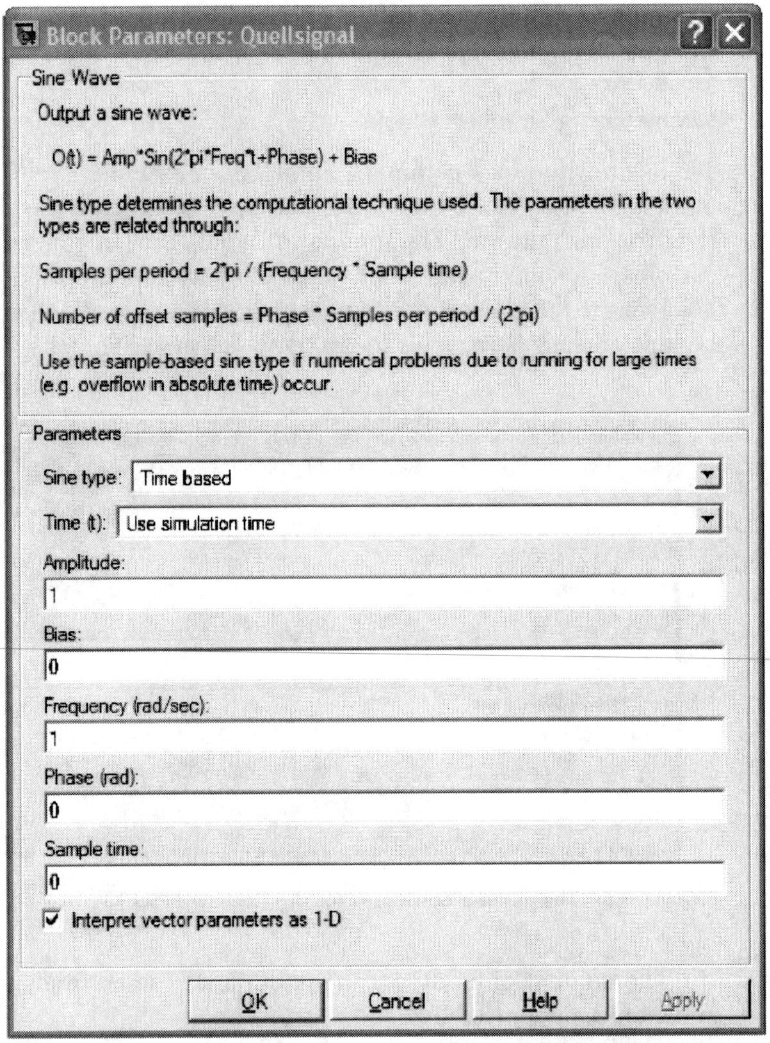

FIGURE 2.7   **The parameter window for the Simulink block** Sine Wave (Source signal).

Here, we can set the amplitude, frequency (in rad/s), and phase. We leave the parameter Sample time at 0 since we want to carry out a (quasi-) continuous simulation. It should only be changed if we perform an explicitly discrete (in time) simulation. In that case we would have to enter a sample time in this space.

For the present example we change the parameters to an amplitude of 2 and a frequency of 2π rad/s, corresponding to 1 Hz. The appropriate entry for this is `2*pi` and not, for example, `6.28` or the like, since Simulink, like MATLAB, recognizes the symbol pi and, therefore, its (numerically) "exact" value. Anyway, why put in a faulty value here?

We choose `pi/4` for the phase. The remaining entries are unchanged.

The integrator block calls for, among others, the initial value for the integration (the parameter `Initial condition`). This is especially important when we want to solve differential equations using Simulink, since the initial value must enter into these equations. For the present example, the initial value should be left at the default, 0. The other parameters are relevant to special forms of integrators and are left at the defaults for our example.

The `Scope` block also has to be parametrized. A double-click on the scope block and a click on the `Parameters` icon (on the right next to the printer icon; see Fig. 2.8) opens up a file-card window into which the parameters can be entered. The possibility of having the read signal stored

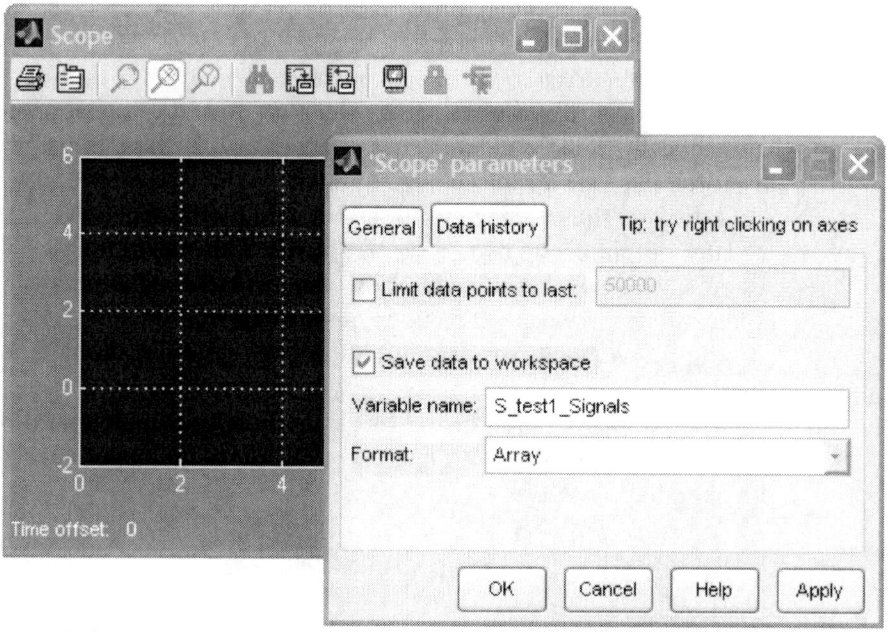

FIGURE 2.8 **Display window for the Simulink block** Scope **with an open** Parameters **window.**

directly as a MATLAB variable is very interesting. Fig. 2.8 shows how the scope signal can be saved in standard MATLAB matrix format[2] as the variable `S_test1_signals`.

Unfortunately, this option (`Array`) is not the default. Thus, you should remember at this point to convert to the matrix format if you want to use this capability for data export to the MATLAB workspace. Also, the check mark next to `Limit data points to last:` should always be removed, since otherwise only the last part (here just the last 5000 points) would be displayed in long simulations.

In other respects, the display of the signal in Scope is set up *after* the simulation using the toolbar buttons. The reader should simply check these out once. The `Scope` should be opened up right away, anyhow, for, unlike in earlier versions of Simulink, the graphics are not opened up automatically after the simulation.

The last bit of polish is provided by the labeling mentioned above. Double-clicking in the model window yields a blinking cursor. Here, free text can be typed into a display label and can be modified according to taste using the menu entry `Format - Font`. In this way the system can be supplied with a title and production date.

The labeling of signal arrows is also very useful. Double-clicking on one of these arrows makes it possible to give this signal a name. These names move with the arrow even if it or an associated block is shifted. Names of inputs to blocks, like that for `Mux`, which are linked to signal names will be automatically changed along with it. If necessary, the block must be enlarged or moved in order for this change to work.

Likewise, the labeling of the system with plot instructions is very useful for later output of the results via MATLAB. This makes it easier to use the system later, since these specifications can just be brought into the MATLAB command window using the cut-paste mechanism.

Fig. 2.9 shows the result of our efforts, the system `s_test1.mdl` ready to go.

The parametrization of the blocks is thereby complete and if we now quickly save our system with `File - Save`, we no longer have to worry if the operating system suddenly crashes.

---

[2]In MATLAB 7 and Simulink 6, as noted in Chapter 1, other data structures that go beyond the matrix format are possible. We recommend using the matrix format unless compelling reasons indicate the contrary.

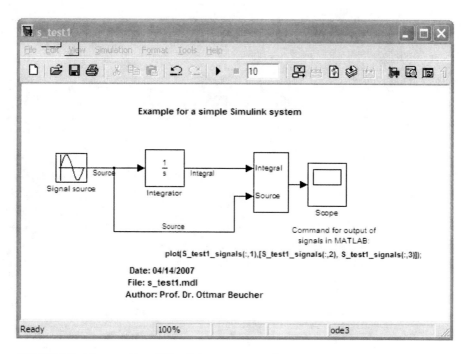

FIGURE 2.9   The completed Simulink system s_test1.

2.2.3 Simulink Simulation

We set the simulation parameters by calling the menu command `Simu-lation - Configuration Parameters`. This action opens up the admirable `Configuration Parameters` window shown in Fig. 2.10.

Normally, the combined options for the solver are displayed on opening the entry `Select: Solver`. A solver is the procedure for *numerical solution of differential equations*, which is to be used for the present simulation The reader might wonder what the model just developed has to do with a differential equation. Hopefully, this will be clearer after reading Section 2.3. But first, refer to Problem 86.

The procedures can be chosen in the pull-down menus `Solver Options Type:` and `Solver Options Solver:`. They are divided into two classes, with and without variable step size adjustment (see the discussion on variable step size below). The class is fixed in the pull-down menu `Type` by choosing the parameter `Fixed step` or `Variable step`.

At this point discussing the procedures in detail would take us too far afield, and that is also unnecessary for understanding the following section. A

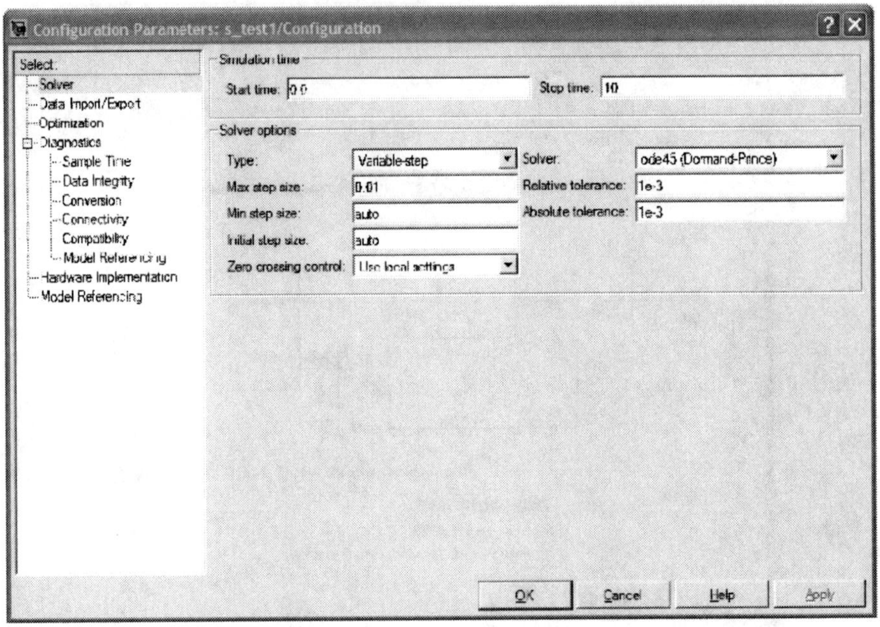

FIGURE 2.10   **The Configuration Parameters (simulation parameter) window.**

more profound discussion of the procedures would go well beyond the scope of this MATLAB-Simulink introduction, for the most modern mathematical procedures are brought into use. At the beginning we only need a few methods `ode23`, `ode45`, and `ode3` which are based on the familiar Runge-Kutta method, which we are already somewhat acquainted from Chapter 1. You can consider the other procedures after you have enough experience with simulations and come upon problems that require their use.

The class of `Variable-step` procedures work with a built-in *step size control*; that is, they change the step size of the numerical solution procedure in accordance with the dynamic behavior of the solution. If the solution varies little, then the step size is automatically set larger, while if the solution varies a lot, it is iterated with smaller step sizes. How much these step sizes are able to vary and with what tolerance can be controlled using the parameters `Step size` and `tolerance`. To go into more detail here would take us too far. The default values are sufficient for most cases.

If you select the `Fixed-step` procedure class, you'll see that the menu for the step width changes. Now the desired (and thence, fixed) step size can be specified for the procedure. The other settings should initially be left unchanged.

The parameters `Start time` and `Stop time` are self-explanatory.

Of the further options that can be chosen under `Select:`, only the entry `Data Import/Export` is of interest in the beginning. Under the heading `Save Options` the solution can also be interpolated between the chosen points in the variable step size procedure, if needed, in order to produce a smoother graphical solution. By selecting `Produce specified output only` in the `Output Options` it can be specified exactly at which points the solution is to be calculated. These points will be specified concretely in a suitable time vector (e.g., `(0:0.1:10)`) under the parameter `Output Times` that appears then. Internally, of course, the procedure calculates with variable control as usual and the solution is then determined by interpolation before output in the desired places. This way of processing the solution is particularly useful during operation with variable step sizes (perhaps for an optimal simulation time) when different solutions have to be compared with one another. In this case the solutions generally have different reference points, which can make comparison very difficult in certain cases. By choosing `Produce specified output only`, however, it is possible to *force* all the simulations to provide values at the same reference points.

But in the present example we initially make it easy and *force* Simulink from the start to employ a fixed step size by selecting the fixed-step procedure `ode3` and setting the `step size` parameter to `0.01`.

The step size setting has a direct effect on the *duration of the simulation*, which can become unacceptably long with step size values that are too small. Here, if necessary, a compromise between the duration and precision of the simulation can be found by testing with multiple experiments. Alternatively, a `variable step` procedure can (and should) be used.

After the simulation parameters have been set, the simulation can be started with the menu call `Simulation - start`. Alternatively, the triangle symbol in the icon toolbar can be clicked. Then the sinusoidal signal and associated integral shown in Fig. 2.11 will be plotted in the `Scope` block.

Since we have stored the result of the scope output as a MATLAB variable, `S_test1_signals`, we can also display the graph in MATLAB. A look at the workspace with

```
>> whos
  Name                  Size              Bytes  Class

  S_test1_signals       1001x3            24024  double array
  t                     1001x1             8008  double array

Grand total is 4004 elements using 32032 bytes
```

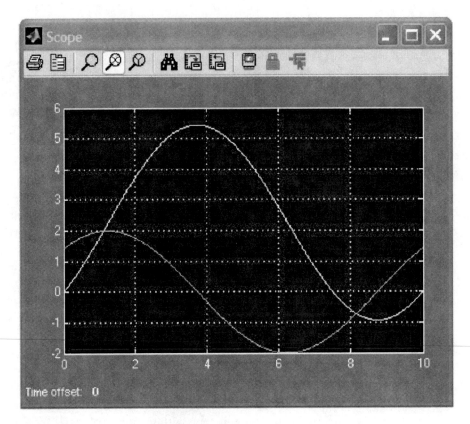

FIGURE 2.11   **The result of the sample simulation.**

shows that *three* vectors are saved as columns. One of the columns, the first, is the time vector (vector of the reference points for the solutions). In the present case the time vector is also supplied to the MATLAB workspace as the variable **t**, since the option **Time** was selected in the **Configuration Parameters** window under **Data Import/Export - Save to Workspace**.

```
>> plot(S_test1_signals(:,1), ...
        [S_test1_signals(:,2),S_test1_signals(:,3)])
>> title('Result of s_test1 with ode3')
>> xlabel('Time /s')
>> ylabel('Function value')
>> grid
```

The result can also be produced with labels in MATLAB (Fig. 2.12).

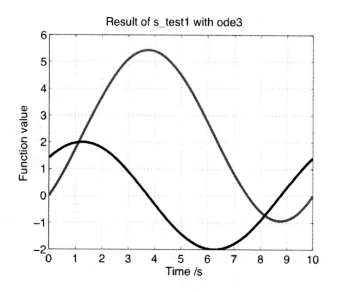

FIGURE 2.12 **The result of the sample simulation after processing by MATLAB.**

PROBLEMS

Work through the following problems to practice using Simulink.

NOTE Solutions to all problems can be found in Chapter 4.

Problem 85
Try out the test system **s_test1** with different simulation step sizes and using step size control.

Compare, in particular, the calculation time for a step size of 0.00001, once using **ode3** directly with this step size and again using **ode23** with conversion of the answer to a step size of 0.00001. Interpret the results.

Problem 86
Think about why the result of the simulation of the test system **s_test1** displays the *solution of the differential equation* (*initial value problem*)

$$\dot{y}(t) = x(t), \quad y(0) = 0. \tag{2.1}$$

Which of the signals is $x(t)$ and which is $y(t)$?

Problem 87
Design a Simulink test system **s_soldiff** for the differentiator block **Derivative**. For this it is best to modify the system **s_test1**.

Next, experiment with this as in Problem 85.

Problem 88

Think about how one could solve the initial value problem

$$\dot{u}(t) = -2 \cdot u(t), \quad u(0) = 1 \tag{2.2}$$

using Simulink and with the aid of the **Integrator** block and set up a Simulink system of this sort.

Compare the resulting numerical solution with the exact solution.

2.3 SOLVING DIFFERENTIAL EQUATIONS WITH SIMULINK

The possibility of solving differential equations with MATLAB has already been mentioned in Chapter 1, Section 1.6.6.

Solving more complicated nonlinear equations is much simpler with Simulink than shown there. The trick is to convert the differential equation into a dynamic system, which can be portrayed in the form of a block diagram in Simulink. In this section we shall clarify the procedure for this conversion using a few examples.

A Simple Example

To warm up, we begin with a simple second order initial value problem. We shall solve the differential equation

$$\ddot{y}(t) = -y(t), \qquad y(0) = 1, \dot{y}(0) = 0. \tag{2.3}$$

The solution of this differential equation is simple, just

$$y(t) = \cos(t). \tag{2.4}$$

Basically, here we should note that when a new topic or a new concept is being introduced, setting up *simple examples* is a reasonable strategy. This is especially true when becoming familiar with new software. The *idea* for solving Eq. (2.3) with Simulink is the following: by systematic integration of the derivative of the desired solution with the Simulink integrator, one obtains the solution. Suppose for a moment that $\ddot{y}(t)$ is already *known*; then the solution $y(t)$ can be calculated using the integrator chain shown in Fig. 2.13.

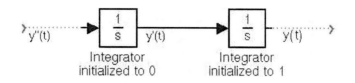

FIGURE 2.13   The technique for integrating to find y(t).

Of course, the integrators have to be initialized to correspond to the initial values; thus, in the present case the first integrator is set to 0, since the value of the output signal $\dot{y}(t)$ at time 0 (the beginning of the simulation) must indeed be 0 according to the initial conditions. The second integrator is set to 1, since the value of the output signal $y(t)$ should be 1 at time 0 according to the initial conditions.

So far, so good, if only we had $\ddot{y}(t)$!

The esteemed reader will now ask in confusion: where can you get it without stealing it? The differential equation, which we have only honored with a fleeting glance up to now, gives us some friendly information. It says that $\ddot{y}(t)$ is simply $-y(t)$.

We simply *connect* the output of the second integrator to the input of the first, but without forgetting the negative sign. At first glance this seems phony, in that we really want to calculate $y(t)$. How can we rely on $y(t)$ for determining $\ddot{y}(t)$?

The resolution of this apparent contradiction lies in the way numerical procedures iteratively calculate the values of the solution $y(t)$. Starting with the *known* initial value $y(0)$ the other reference points are approximately determined. Thus, at the beginning of the simulation a value for the first integrator block (namely $-y(0)$) is available and, therefore, in all the succeeding iteration steps as well.

So much for the operation in principle. We can confidently leave the details to Simulink's numerical solution procedures for initial value problems. Let us consider the Simulink system (**s_de2or**) shown in Fig. 2.14.

This system is enhanced with a graphical output and a block for linking the result to the MATLAB workspace.

With the parameter set shown in Fig. 2.15, the simulation yields the cosine predicted in Eq. (2.4).

Naturally, the quality of the numerical solution depends on the chosen parameters. The reader can make this clear by running the simulation a couple of times with different step sizes.

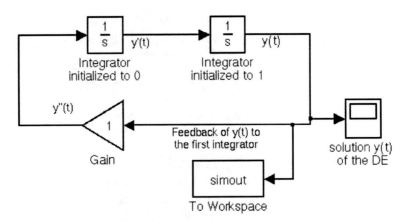

FIGURE 2.14   The Simulink system for solving Eq. (2.3).

Example: The Logistic Differential Equation

In the next example, we examine a famous equation from the theory of growth processes, the so-called *logistic differential equation*:

$$\dot{P}(t) = \gamma P(t) - \tau P^2(t) . \tag{2.5}$$

Here, $P(t)$ refers to a population of individuals at time t and γ is the growth rate per unit time, while τ is the loss rate. In order to ensure that the loss rate has a greater effect on the overall growth behavior in very large populations (a reasonable assumption, given a lack of food, etc.), the population is quadratic in the loss term.

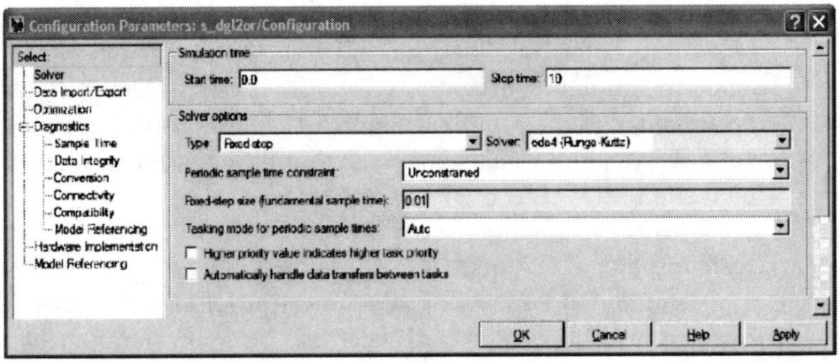

FIGURE 2.15   The parameter window for the Simulink system of Fig. 2.14.

In order to solve this differential equation for an initial value of $P(0) = 10000$ individuals, a growth rate of $\gamma = 0.05$ and a loss rate of $\tau = 0.0000025$ with Simulink, we use an *integrator block* whose input is $\dot{P}(t)$ and whose output is $P(t)$ and which is initialized to the initial value 10000.

According to Eq. (2.5) $\gamma P(t) - \tau P^2(t)$ is fed to the input of the integrator, which can be done from the integrator output by feedback.

This leads to the Simulink system (file **s_logde.mdl** in the accompanying software) shown in Fig. 2.16.

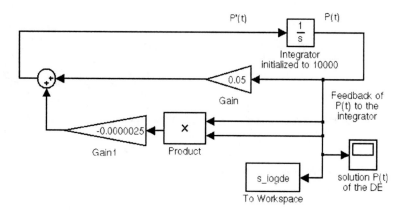

FIGURE 2.16 The Simulink system **s_logde** for solving a logistic differential equation.

A fixed step size of 1 (the unit of time) yields the solution shown in Fig. 2.17.

This agrees well with the theoretical solution

$$P(t) = \frac{\gamma}{\tau + \gamma C e^{-\gamma t}}, \quad C = \frac{1}{P(0)} - \frac{\tau}{\gamma}, \tag{2.6}$$

for the chosen value of γ and τ. It is interesting to point out here that the simulation is considerably faster with automatic step size control. Of course, for a correct interpretation of the result the internally generated time vector must also be saved.

Example: Mechanical Oscillations

As a final example, let us consider an oscillating mechanical system, represented by a mass m kg, a spring with spring constant c N/m, and a medium that damps the oscillation, represented by a coefficient of friction d Ns/m.

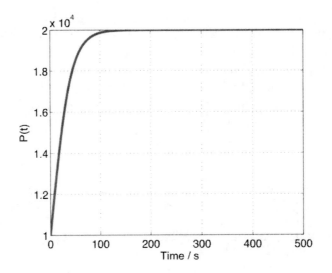

FIGURE 2.17   A Simulink solution of the logistic differential equation.

The corresponding second order differential equation for free damped oscillation is

$$m\ddot{x}(t) + d\dot{x}(t) + cx(t) = 0 \,. \tag{2.7}$$

Here $x(t)$ represents the deviation from the equilibrium position at time t.

In gases or liquids and for rapid motion of masses, it is found definitively, however, that the force owing to friction on an object is generally not proportional to the velocity $\dot{x}(t)$, as in Eq. (2.7), but is *proportional to the square of the velocity* (newtonian friction).

This leads to the following differential equation:

$$m\ddot{x}(t) + b\dot{x}^2(t) + cx(t) = 0 \,, \quad \text{if } \dot{x}(t) \geq 0 \,, \tag{2.8}$$
$$m\ddot{x}(t) - b\dot{x}^2(t) + cx(t) = 0 \,, \quad \text{if } \dot{x}(t) < 0 \,. \tag{2.9}$$

The equation is split into two because the reversal of the direction of the velocity has to be taken into account. The sign in front of $\dot{x}(t)$ is lost when the velocity is squared, and this sign indication has to be restored.

Eq. (2.8) can be rewritten using the so-called *sign function*,

$$sgn(y) := \begin{cases} 1 & \text{for } y \geq 0 \,, \\ -1 & \text{for } y < 0 \end{cases}$$

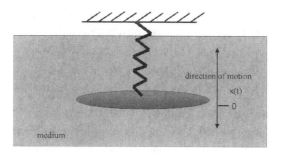

FIGURE 2.18   **Mechanical oscillator (steel plate).**

as follows:

$$m\ddot{x}(t) + b \cdot sgn(\dot{x})\dot{x}^2(t) + cx(t) = 0 . \qquad (2.10)$$

We now attempt to solve this equation using Simulink and construct a suitable system for his purpose.

First, we require numerical values for the quantities involved. To do this we assume that the oscillating mass is a round steel plate (Fig. 2.18) with $m = 0.5$ kg. The area A of the plate lies perpendicular to the direction of motion. The coefficient of friction can be assumed to have the form

$$b = c_w \frac{1}{2}\varrho A \qquad (2.11)$$

where ϱ is the density of the medium.

c_w is the so-called drag coefficient. It depends on shape of the object. For a plate-oriented perpendicular to the direction of motion, as we have assumed in our model, c_w can be taken to be between 1.1 and 1.3. For simplicity, in the following example we take $c_w = 1$ in order to avoid having to carry the value of c_w along in the calculations.

The mass is given by the volume cm of the plate times the density, $h \cdot \pi r^2$ cm^3, where h is the thickness of the plate and r is its radius (all lengths in cm). A density of 7.85 g/cm^3 can be assumed for unalloyed steel. Then the mass is

$$500 = 7.85 \cdot h \cdot \pi r^2 ,$$

for a thickness of $h = 1$ cm and a radius of 4.5027 cm (i.e., a cross-sectional area of 63.6943 cm^2).

The gas in which the disk is suspended is *air*, with a density of $\varrho = 1.29$ kg/m^3, so that b has the numerical value[3]

$$b = \frac{1}{2} \cdot 63.6943 \cdot 10^{-4} \cdot 1.29 \, \text{m}^2 \frac{\text{kg}}{\text{m}^3} = 0.00411 \, \frac{\text{kg}}{\text{m}} \, .$$

Here, we set a value of 0.1552 N/mm for the spring constant, which corresponds to 155.2 N/m. This yields the following differential equation:

$$0.5 \cdot \ddot{x}(t) + 0.00411 \cdot sgn(\dot{x}) \cdot \dot{x}^2(t) + 155.2 \cdot x(t) = 0 \, . \tag{2.12}$$

Let us review the physical units once again at this point. The terms in Eq. (2.12) are *forces*. All the units must, therefore, reduce to N $= \frac{\text{kg} \cdot \text{m}}{\text{s}^2}$. In the first term, the units of the 0.5 are kg and those of the second derivative of the displacement $x(t)$ with respect to time are $\frac{\text{m}}{\text{s}^2}$; in the second term b is in units of $\frac{\text{kg}}{\text{m}}$ and $\dot{x}^2(t)$ is in units of $\frac{\text{m}^2}{\text{s}^2}$. sgn$(\dot{x})$ is dimensionless. In the third term, c is in units of N/m and $x(t)$ is in units of m. All three terms are, therefore, in units of N.

Now we can proceed in good conscience to setting up a numerical solution with Simulink.

Since we are dealing with a second order equation, we require *two integrators* which, subject to the initial conditions represented in the initialization of the integrators, integrate $\ddot{x}(t)$ to obtain the solution $x(t)$. As initial conditions we take a displacement of $x(0) = 1$ m and an initial velocity of $\dot{x}(0) = 0$ m/s. We feed the (negative) sum of $x(t)$ and $\dot{x}^2(t)$ with the coefficients given in Eq. (2.12) and the nonlinearity sgn$(\dot{x})$ back into the first integrator.

This yields the Simulink system **s_denon** shown in Fig. 2.19.

In order to be able to display the solution with step size control activated in MATLAB, besides directing the solution into a **scope** sink, we direct it into a MATLAB sink, which will store the solution in a vector **DEsolution**. In order to have the time reference, we designate the parameter **time** in the entry **Data Import/Export** under **Simulation Configuration Parameters**. The default name **t** for the time vector can be changed at will. In both cases, you have to make sure to set **array** for the prescribed places as the save format, so that the results appear as matrices and vectors in the workspace.

---

[3]It is fated that the physical units must be brought into a form in which SI units are finally used. Otherwise, there is a danger of using incorrect numerical values, such as when mm and then later m is used in the calculations.

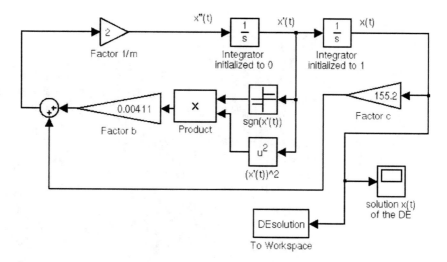

FIGURE 2.19   A Simulink system for solving Eq. (2.12).

The solution for the set parameters and initial values is displayed in Fig. 2.20. It was calculated using the `ode45` procedure with step size control activated (parameters:`Initial Step Size = auto`, `Max Step Size = 10`, `tolerances = 1e-3`)) over the time interval [0, 10].

PROBLEMS

Work through the following problems to practice the techniques for numerically solving differential equations with Simulink.

NOTE ▶ Solutions to all problems can be found in Chapter 4.

Problem 89

Model the mechanical oscillator problem with a hemisphere instead of a steel plate and water as the medium instead of air. The drag coefficient c_w varies for a hemisphere depending on whether its flat or round side lies in the direction of the motion. For the flat side, it is $c_w = 1.2$ and for the round side, $c_w = 0.4$. Take this into account in a modified Simulink system where you distinguish the two cases using a `switch`-Blocks block.

Problem 90

Design and test a Simulink system for solving Eq. (1.4) for the mathematical pendulum.

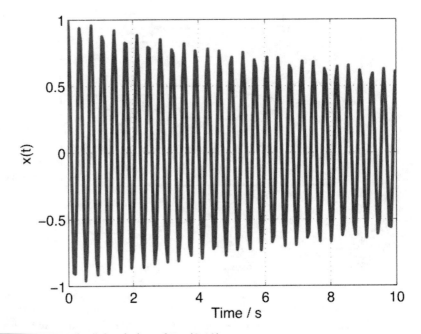

FIGURE 2.20 Simulink solution of Eq. (2.12).

Problem 91
Design and test a Simulink system for solving the initial value problem

$$\ddot{y}(t) + y(t) = 0, \quad y(0) = 1, \dot{y}(0) = 0. \tag{2.13}$$

Compare the result with the exact solution. Next, incorporate the capability of simulating a perturbation function (i.e., an inhomogeneity, with the right-hand side $\neq 0$) into the Simulink system. Try out the system with the perturbation function e^{-t}.

Problem 92
Solve the initial value problem

$$t\ddot{y}(t) + 2\dot{y}(t) + 4y(t) = 4, \quad y(1) = 1, \, \dot{y}(1) = 1 \tag{2.14}$$

with a suitable Simulink system.

Problem 93

Solve the system of differential equations

$$\dot{y}_1(t) = -3y_1(t) - 2y_2(t), \quad y_1(0) = 1, \tag{2.15}$$

$$\dot{y}_2(t) = 4y_1(t) + 2y_2(t), \quad y_2(0) = 1 \tag{2.16}$$

with a suitable Simulink system. Compare the solution with the exact solution, which you can calculate by hand or with the aid of the symbolics toolbox.

2.4   SIMPLIFICATION OF SIMULINK SYSTEMS

The last examples in the preceding section, as well as those in Problems 89 to 93, show that Simulink systems for solving differential equations (and, therefore, for simulating dynamic systems) can rapidly come to contain quite a few blocks even for comparatively small problems. This is especially true for the problems that arise in industrial applications.

For these simple examples, this effect is naturally first traceable to the fact that the corresponding blocks were used for even the lowest level operations, such as addition or scaling. This does, indeed, contribute to traceability of the equations within the block diagram, but quickly makes it downright unintelligible. As noted in Section 2.2, subsystems (partial systems) can, in general, be assembled into their own Simulink blocks for simplifying Simulink systems. The resulting hierarchical arrangement of the problem corresponds to the modular arrangement by functions in MATLAB programs. This sort of modular arrangement is indispensable for most practical problems. The corresponding technique is discussed briefly in this section.

2.4.1   The Fcn Block

It should be pointed out first that Simulink systems can often be considerably simplified through clever use of the **Fcn** block from the block library **UserDefined Functions**. With this block it is possible to assemble *entire formulas* into a unit, so that the lowest level elementary blocks (e.g., **Sum** or **Gain**) can be removed (see the solutions to Problems 89 and 90). Let us clarify this with the example of the system in Fig. 2.16 for solving the logistic differential equation. Here, the entire right-hand side of Eq. (2.5) can be combined using the **Fcn** Block.

 The result (file **s_logde2.mdl** in the accompanying software) is displayed in Fig. 2.21.

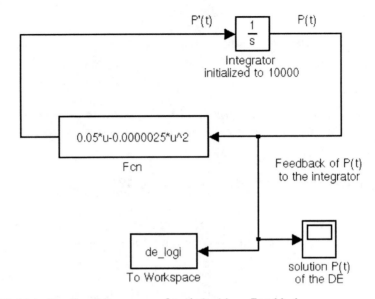

FIGURE 2.21   The Simulink system `s_logde2` with an `Fcn` block.

You can see that the whole feedback branch of Fig. 2.16, consisting of two **Gain** blocks, a summation block, and a multiplication block, has been assembled into a single block.

When using the **Fcn** block it should be kept in mind that the input signal for the block is always called **u**, regardless of how it may be denoted in the system. The input signal can be a *scalar* or *vector* quantity. If multiple signals have to be fed into an **Fcn** block, then they can (and must) be combined beforehand into a *vector* signal using a **Mux** block. With vector input signals, the components are addressed by indexing (**u(1)**, **u(2)**, **...**) within the block. The output signal is always a scalar quantity.

2.4.2   Construction of Subsystems

As noted above, the possibilities for "cleaning up" a Simulink system using **Fcn** blocks are limited in large scale problems.

In that case, modularizing the problem using self-defined Simulink blocks is the appropriate means for bringing order to the chaos. We now clarify this with a small example.

Let us again consider the example of the logistic differential equation. Instead of seizing on the partial system which joins the signals $P(t)$ and $P'(t)$ in

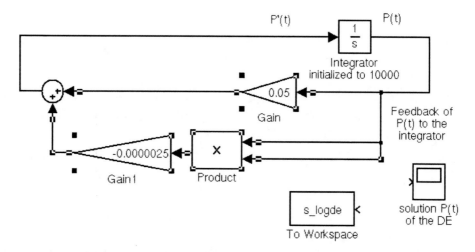

FIGURE 2.22   Selecting the blocks that will be assembled into a subsystem.

Fig. 2.16 as a formula and stuffing it into an `Fcn` block, as we did in Fig. 2.21, it could actually be viewed as a *partial system*, which can be represented[4] as an independent Simulink block with an input and an output.

To do this you first select the system blocks that are to be combined into a subsystem. This can be done most easily by drawing a box with the mouse or, if the blocks do not lie in a rectangle, by clicking with the shift key pressed.

Fig. 2.22 shows this for the system `s_logde`.

Here, it should be kept in mind that besides the blocks, the signal lines have to be selected,[5] especially those for the in- and output signals.

After this selection, you choose the menu entry `Edit - Create subsystem` and get the system shown in Fig. 2.23.

You can see that the marked blocks have been replaced by a single block with an input and output. Its "insides" can be viewed by double-clicking on the newly created block (Fig. 2.24). Depending on the Simulink settings,[6] this will open up a separate model window or display the subsystem in the same window.

---

[4]Of course, this would not normally be done for such a simple problem. The previously proposed solutions are preferable. This example serves only to illustrate the procedure for constructing subsystems.
[5]The sink blocks are initially uncoupled in `s_logde.mdl`, since otherwise three lines would be interpreted as an input signal during subsystem construction.
[6]See the MathWorks *Simulink 6 Handbook*, 2004.

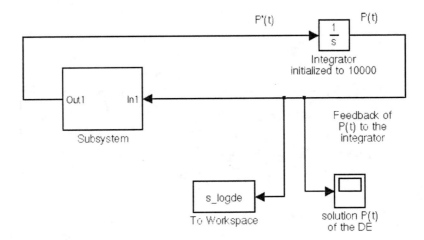

FIGURE 2.23   The system s_logde after creation of the subsystem (file s_Logde3.m).

You can see that two new blocks have been added to the combined blocks. These are the blocks **In1** and **Out1** from the block library's **Sources** and **Sinks**, respectively. The input and output are correspondingly connected to the created Simulink subsystem by means of these blocks.

The system **s_Logde3** is now fully functional and can be executed. Parameters can be changed inside the subsystem by opening the parameter windows of the blocks as usual.

Starting up the so-called **model browser** with the menu command **View - Model Browser Options - Model Browser** displays the hierarchical structure of the system as shown in Fig. 2.25.

The left panel shows a model-tree (which is, of course, very small here) that exhibits the partial systems of which the whole system is composed.

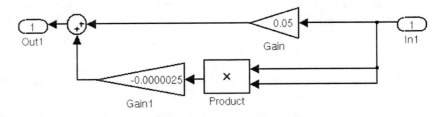

FIGURE 2.24   Structure of the subsystem s_Logde3.m.

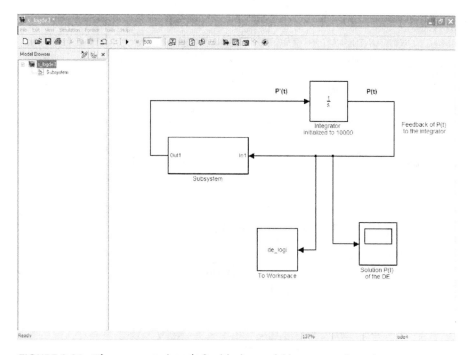

FIGURE 2.25   **The system** s_Logde3 **with the model browser activated.**

With this resource it is very easy to find your way in complicated, nested systems.

As an alternative to the above procedure, a subsystem can also be constructed by selecting the block **Subsystem** from the block library **Ports&Subsystems** and displaying it in a model window. Double clicking on this block opens a model window in which the partial system can be made available.

Finally, it should be mentioned that the self-created blocks can be equipped with extended functionality by means of so-called *masking*. Thus, the blocks can, for example, be provided with a pictogram or, more importantly, the blocks can be set up with their own parameter window, which can forward the parameters on to the underlying blocks. In that way it is no longer necessary to open the subsystems in order to modify parameters.

Afterward, the self-provided blocks can be assembled into their own block libraries. These block libraries can then be used like the original Simulink *block libraries*.

Further discussion of these and other capabilities is not possible in this introduction. The interested reader is referred to the handbooks or to MATLAB help.

PROBLEMS

NOTE   Solutions to all problems can be found in Chapter 4.

Problem 94
Simplify the Simulink system shown in Fig. 2.19 for solving Eq. (2.12) for nonlinear oscillations using an **Fcn** block.

Problem 95
Design a Simulink system which solves the linear system of differential equations (2.15) using **Fcn** blocks.

Problem 96
Design a Simulink system that simplifies the system shown in Fig. 2.19 using a partial system for the blocks contained in the feedback branch.

2.5   INTERACTION WITH MATLAB

In the last example of Section 2.3, we mentioned the possibility of using *MATLAB sink* to transfer Simulink results into the MATLAB workspace.

2.5.1   Transfer of Variables Between Simulink and MATLAB

The use of MATLAB sinks is by no means the only way to interact with MATLAB. For example, if in Section 2.3 it were possible to vary the parameters m, b, and c in solving Eq. (2.10), or even better, the quantities such as the radius r of the plate or the densities of the materials used that define these parameters, Eq. (2.12) could be solved instead of Eq. (2.10), which was set up with specific parameters.

This is readily done in Simulink if the corresponding quantities are defined in advance in the MATLAB workspace and if the variables rather than the numbers are entered for parametrizing the Simulink blocks.

It is also easy to arrange the *return of the time vector* in the example system of Fig. 2.19, as noted above. For this, you only have to enter a suitable variable (e.g., t or time) in the parameter block for the system under the entry Data Import/Export - Save to workspace. Only certain variables can be returned via Save to workspace; the first of these is the internally generated time vector. We won't discuss the other, likewise internally generated variables that can be returned here. The interested reader is referred to the handbook.[7] We have already dealt with the possibility of passing results on to MATLAB via the Scope block in our discussion of parametrizing the scope in Section 2.2.2. But, the parameters of the simulation parameter window, themselves, *must not be numbers*.

These values can also be defined within MATLAB as variable quantities. Let us now clarify these procedures with the aid of the solution of Eq. (2.10) from Section 2.3.

As variables for the experiment we choose the plate radius r and the density ϱ of the material in which the steel plate moves. The spring and, thereby, the spring constant are the same (i.e., we set $c = 155.2$ N/m). All the other quantities are derived from these. In addition, the simulation time, as well as the step size parameter (Fixed step size) should also be variable.

The *most elegant solution* for this problem is to write a *MATLAB function*, which has the variable parameters as functional parameters and calculates the quantities to be supplied to Simulink. After that, we only need to enter the variable quantities in the modified Simulink system s_denon and to start the simulation. The MATLAB function denonpm for defining the parameters looks like this:

```
function [m, b, c, tstep, stime]= denonpm(r, rho, sz, tm)
%
% Function  denonpm
%
% call:  [m, b, c, tstep, stime]= denonpm(r, rho, sz, tm)
%
% MATLAB function for parametrizing the Simulink system
% s_denon2.mdl for solving the nonlinear differential
% equation for a single mass oscillator
%
% Input data:  r   Plate radius cm
%              rho Density in g/cm3
%              tm  Simulation end time
%              sz  Step size for simulation
```

---

[7] See the MathWorks *Simulink 6 Handbook*, 2004.

```
%                       with constant step size
%
%
% Output data: The parameters of the differential
%   equation and the Simulink simulation parameter window

%   Pass through the Simulink parameter block Parameters

tstep = sz;
stime = tm;

% Spring constant is constant
% (enter formula here later if needed)

c = 155.2;                      % N/m

% Plate thickness is constant
% (enter formula here later if needed)

h = 1;                          % cm

% Calculation of m and b

m = (7.85*h*pi*r^2)/1000;    % mass of the steel plate in kg
b = ((1/2)*rho*pi*r^2)/10;   % Damping constant in kg/m (cW=1)
```

A call with the parameters from Section 2.3, for example, then yields:

```
>> [m, b, c, tstep, stime]= denonpm(4.5027, 1.29/1000, 0.001, 10)

m =

    0.5000

b =

    0.0041

c =

  155.2000

tstep =

  1.0000e-003
```

```
stime =

    10
```

The modified Simulink system, which we save under the name **s_denon2**, can be seen in Fig. 2.26.

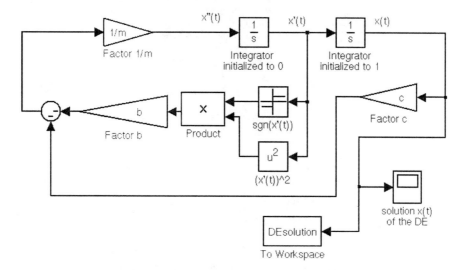

FIGURE 2.26 The Simulink system s_denon2 for solving Eq. (2.12) with common parameters from the MATLAB workspace.

The corresponding parameter window (`Configuration Parameters`) is displayed in Fig. 2.27.

2.5.2 Iteration of Simulink Simulations in MATLAB

In many cases a Simulink simulation depends on a very large number of parameters. The simulation of an oscillator in the preceding Section 2.5.1 is only one of innumerable examples of such simulations. For this reason it is often very useful to vary one or more parameters in order to establish the dependence of the simulated system on this (or these) parameter(s).

Naturally, it would be very inconvenient to start the Simulink system *manually* every time. In MATLAB this process can be *automated*.

Calling Simulink systems via MATLAB

The trick involves (repeatedly) calling the Simulink system via MATLAB or within a MATLAB function.

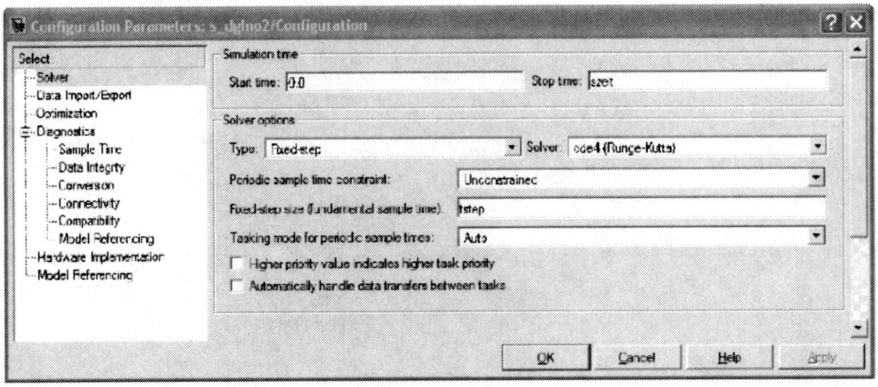

FIGURE 2.27   **The configuration parameters window for Simulink system** s_denon2.

This mechanism can, moreover, be used to call a Simulink simulation from MATLAB, without having to start Simulink explicitly for this purpose. The advantage of a call of this type is a generally *higher speed of execution*.

A Simulink system is called from MATLAB or from a MATLAB program using the MATLAB function **sim**.

Exact information on the various ways **sim** is used can be obtained by entering **help sim** or by taking a look at MATLAB help.

For our (and, indeed, most) purposes it is sufficient to know the important parameters of **sim** and, if necessary, to supply it with values. Let's first examine an extract of the response of MATLAB to **help sim**:

```
>> help sim

 SIM Simulate a Simulink model
    ...


    The SIM command also takes the following parameters.
    By default time, state, and output are saved to the
    specified left-hand side arguments unless OPTIONS
    overrides this. If there are no left hand side
    arguments, then the simulation parameters dialog
    Workspace I/O settings are used to specify what data to log.

    [T,X,Y]          = SIM('model',TIMESPAN,OPTIONS,UT)
    ...


        T                : Returned time vector.
```

```
X            : Returned state in matrix or structure
               format. The state matrix contains
               continuous states followed by discrete
               states.
Y            : Returned output in matrix or structure
               format. For block diagram models this
               contains all root-level outport blocks.

...

'model'      : Name of a block diagram model.
TIMESPAN     : One of:
                   TFinal,
                   [TStart TFinal], or
                   [TStart OutputTimes TFinal].

               OutputTimes are time points which will be
               returned in T, but in general T will include
               additional time points.
OPTIONS      : Optional simulation parameters. This is a
               structure created with SIMSET using name
               value pairs.
UT           : Optional extern input.

...
```

```
Specifying any right-hand side argument to SIM as the empty
matrix, [], will cause the default for the argument to be used.

Only the first parameter is required. All defaults will be
taken from the block diagram, including unspecified options.
Any optional arguments specified will override the settings
in the block diagram.

See also sldebug, simset.
```

The main point is:

sim is called with the *name of the Simulink model* and, if necessary, with a vector (**TIMESPAN**) as parameters; this sets the discrete time points for the iteration. If desired, the time vector can be returned through the parameter **T** as an output vector, likewise the so-called state variable **X** of the simulation (further discussion is beyond the scope of this introduction but we won't need it here) and the output variables **Y**, which represent the results of the simulation in general. These can, of course, only be returned under certain conditions, namely, if these variables are connected with an **Outport** block (see below) in the Simulink model.

All the remaining parameters that are not addressed directly will be taken over as they are entered in the definition of the Simulink system in the parameter window.

In principle, a mix of settings in the system itself and in settings via the parameter lists of **sim** is to be recommended.

MATLAB, of course, also provides the capability of changing all the parameters in the parameter window through the parameter lists of **sim**. This is accomplished by the function **simset**. For example, this can be used to set the numerical solution procedures (integration procedures) or tolerances and step sizes. Since we are primarily concerned in the following with modifying Simulink systems that have fixed settings but varying internal block parameters, which are controlled via the MATLAB interface, we will not discuss these possibilities here.

For our purposes, these considerations lead to the following call syntax for the function **sim**:

```
[t,x,y] = sim ('system', [starttime, endtime]);
```

All other parameters should be set via the Simulink parameter window.

The parameter `'system'` is, as noted above, the name of the Simulink system to be executed (this is a string, so it has to be enclosed in `' '`). For example, if the system from Fig. 2.26 is to be simulated, then `'s_denon2'` is entered.

The parameter vector `[starttime, endtime]` marks the start and end times for the simulation. The return vector `[t,x,y]` appears in exactly the same sequence in the Simulink parameter window under the entry **Data Import/Export - Save to workspace**. As mentioned above, *output ports* have to be provided in the Simulink system for the parameter **y**. The output port appears under the name **Out1** in the block library **Sinks**. Each output signal must pass into a port of this type (see Section 2.4). A column is set up in **y** for each port.

Before we proceed to the proposed iteration scheme, let us illustrate the call for a Simulink system from MATLAB with an example:

```
>> [t,x,y] = sim ('s_denon', [0, 10]);
```

In this call, all the parameters up to the start and end time points for the simulation remain unchanged from their settings in the parameter window for the system **s_denon** when last used; in the present case, we used the **ode45** procedure with the default tolerances for the step size control.

Information on the return vectors and matrices is provided by calling whos:

```
>> whos
  Name              Size          Bytes  Class

  DEsolution        212x1          1696  double array
  t                 212x1          1696  double array
  x                 212x2          3392  double array
  y                 0x0               0  double array

Grand total is 848 elements using 6784 bytes
```

The time vector and (via MATLAB sinks) the solution DEsolution are returned. It is interesting that the vector y is empty. This happens because no output port is available.

Setting Simulink Parameters with set_param

The system s_denon is still not suitable for an iterated simulation. First of all, the Scope block interferes, as it would be called repeatedly during iteration and the simulation unnecessarily slowed down. Furthermore, the MATLAB sinks are superfluous when we are using the output parameters t and y.

A third problem is associated with the MATLAB function set_param, which will be described below. It sets parameters of a Simulink system from MATLAB and uses the block name, among others, for this purpose. Unfortunately, however, the function has an allergic reaction to block names with special characters, such as we have used in the system s_denon (e.g., the factor 1/m). The block names, therefore, must not contain any special characters (e.g., \ or line breaks).

In order to deal with all these aspects, in the following we change the system s_denon to a new system s_denon3, in which these points are taken into account and write a MATLAB program which can call this system repeatedly.

Fig. 2.28 shows the system s_denon3.

You can see that the MATLAB sink and the Scope block have been replaced by an output port. This corresponds to the output vector[8] y. In addition, the names of blocks whose parameters are to be set from outside are replaced by names without any special characters.

---

[8]As noted above, with multiple ports there is an output *matrix*.

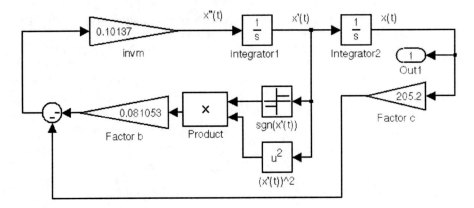

FIGURE 2.28 The Simulink system s_denon3.

As mentioned above, the MATLAB function **set_param** is responsible for setting the parameters out of MATLAB. In order to understand how it operates, consider the following extract from MATLAB help:

```
>> help set_param

SET_PARAM Set Simulink system and block parameters.
    SET_PARAM('OBJ','PARAMETER1',VALUE1,'PARAMETER2',VALUE2,...),
    where 'OBJ' is a system or block path name, sets the specified
    parameters to the specified values.  Case is ignored for
    parameter names.  Value strings are case sensitive.  Any
    parameters that correspond to dialog box entries have
    string values.

Examples:

    set_param('vdp','Solver','ode15s','StopTime','3000')

sets the Solver and StopTime parameters of the vdp system.

    set_param('vdp/Mu','Gain','1000')

sets the Gain of block Mu in the vdp system to 1000 (stiff).

    set_param('vdp/Fcn','Position','[50 100 110 120]')

    ....
```

The function `set_param` can thus define anew from outside certain parameters of a system whose name appears as a first parameter. A characteristic of this function is that *only strings* (text) serve as parameters. Thus, in particular, besides the names of the block parameters (e.g., `'gain'`), the new value `'1000'` is passed on as a string, making it necessary to convert the parameters into text in advance.

Once the syntax of the function `set_param` has been clarified thus far, the question remains of where the parameter names of a Simulink system come from. The simplest way to get the parameters of a block is to drag the mouse pointer over the block. As noted above, a window then opens up (with a bit of a delay) in which the important parameters and their current values are displayed. But this method only yields a few of the parameters of the entire system.

A more precise view into the world of the parameters of the Simulink system is possible with the function `get_param`.

With the command

```
>> FactorcPars = get_param('s_denon3/Factorc','DialogParameters')

FactorcPars =

                      Gain: [1x1 struct]
            Multiplication: [1x1 struct]
            ....
```

for example, you can discover which of the so-called *dialog parameters* the block has. These are the parameters that can be set in the parametrizing window of the block, in the present case the gain block **Factorc** of the Simulink system **s_denon3**.

The parameters are returned in a structure whose name entry is identical with the name of the parameter. The values of the parameters can likewise be obtained with `get_param`. Thus, the call

```
>> currentGain = get_param('s_denon3/Factorc','Gain')

currentGain =

155.2
```

shows, for example, that the parameter **Gain** of the block named **Factorc** is currently set at the value **155.2**.

With

```
>> Integrator1Pars = ...
      get_param('s_denon3/Integrator1','DialogParameters')

Integrator1Pars =

       ...
       InitialCondition: [1x1 struct]
       ...
```

and

```
>> currentInCon = ...
      get_param('s_denon3/Integrator1','InitialCondition')

currentInCon =

0
```

we find that the integrator block **Integrator1** has a parameter **InitialCondition**, which currently is **0**.

Not only the parameters of a block, but also the whole system can be obtained in a similar way. In the present example, this is done using

```
SystemParams = get_param('s_denon3','ObjectParameters')

>> SystemParams =

             Name: [1x1 struct]
              Tag: [1x1 struct]
      Description: [1x1 struct]
             Type: [1x1 struct]
             ...
```

If the parameter names have been obtained in this way, then, as we have already explained, they can be changed from outside using the MATLAB command **set_param**. The two examples of parameters, for example, can be redefined as follows:

```
>> set_param('s_denon3/Factorc','Gain','200');
>> set_param('s_denon3/Integrator1','InitialCondition','1');
```

It should be pointed out once again that with the **set_param** command (as with the **get_param**) the parameters all have to be set in apostrophes, since they must be strings.

After these calls, the parameters are also changed visually in the system s_denon3, as you can easily see.

The system can then, as indicated above, be run through a new simulation with

```
>> [t,x,y] = sim ('s_denon3', [0, 10]);
```

Multiple Calls of Simulink Systems

For repeated calls of this system we use the following MATLAB function denonit, which we have obtained by modifying the function denonpm:

```
function [t,Y] = denonit(r, rho, Fc, tm, step, anfs)
%
% ...
%
% MATLAB function for parametrizing and ITERATIVE execution
% of the Simulink system s_denon3.mdl for solving the nonlinear
% differential equation for a single-mass oscillator.
% The radius r of the oscillating plate will be varied.
%
% input data:  r      Plate radius VECTOR in cm
%                     for each simulation
%              rho    Density rho of the medium in g/cm3
%                     for each simulation
%              Fc     spring constant c in N/m
%                     for each simulation
%
%              Note: the length of the vector r determines the
%                     number of iterations.
%                     In iteration step k, the simulation will
%                     be run with the radius-dependent
%                     mass-parameter m=M(k).
%
%              tm     Simulation end time tm
%                     (the same for all simulations.
%                     Simulations always begin at 0)
%
%              step   Simulation parameter for the step
%                     size in the Simulink parameter window.
%                     It will be processed without step
%                     size control, since all the result
%                     vectors have the same length.
%
```

```
%              anfs   Initial value VECTOR for the
%                     differential equation.
%
% Note: in the present version, NO plausibility tests for
%       the parameters are attempted.
%       No erroneous inputs will be caught.
%       s_denon3.mdl must be set up in a
%       FIXED-STEP procedure.
%
% Output data: t     the time vector
%
%              Y      Matrix of results. For each
%                     iteration a NEW COLUMN is produced.
%


% Calculation of m and b from the input parameters
% First the vectors M and B are constructed; these
% contain the parameters m and b for the given iteration
% as components (see the program
% denonpm about this)

h = 1;                          % cm (fixed thickness)

M = (7.85*h*pi*r.^2)/1000;    % masses of the steel plates in kg
B = ((1/2)*rho.*pi.*r.^2)/10; % damping factor in kg/m

% Defining the initial conditions for the integrators
% along with the step size and the simulation time
% using the function set_param

a0 = num2str(anfs(1));
set_param('s_denon3/Integrator1','InitialCondition',a0);
a1 = num2str(anfs(2));
set_param('s_denon3/Integrator2','InitialCondition',a1);
stepsize = num2str(step);
set_param('s_denon3','FixedStep',stepsize);

% Initializing the output matrix Y as empty
% and the iteration duration

Y = [ ];
iterations = length(r);
```

```
% Calling the iteration loop for the simulation
% (running the iteration by calling \verb"sim" and
% the fixed step size from 0.0 to tm)
% In this version of the program it is necessary to
% check that a fixed-step procedure has been
% set in the parameter window.

for i=1:iterations
    % Set the block parameters with set_param
    % for each new iteration.
    reciprocalmass = num2str(1/M(i));
    set_param('s_denon3/invm','Gain',reciprocalmass);
    b = num2str(B(i));
    set_param('s_denon3/Factorb','Gain',b);
    c = num2str(Fc);
    set_param('s_denon3/Factorc','Gain',c);
    [t,x,y] = sim('s_denon3', [0,tm]);
    Y = [Y,y];
end;
```

This is an extensive set of comments in the function. The initial commentary describes all the important properties that this function should have. The reader should look through these comments carefully since they reveal a lot about the handling of iterated calls in Simulink systems.

Essentially, this function does the following: in each iteration step it sets the parameters of the system anew using a **for**-loop and calls the system using **sim**. The results are entered *column-by-column* in a matrix Y. The time vector **t** is always the same, since a fixed-step procedure is being used.

Compared to previous calls of **set_param**, here we have a difficulty in that the parameters are supplied in the form of *variables*, rather than numbers. The variables can no longer simply be read in as strings in the parameter list of **set_param**, since here the string must be made up of *values* and not of variable names. The function **num2str** serves to place variable *values* in a string. The call for this function is self-explanatory. If not, then the MATLAB help should eliminate any doubts.

A brief sample call should now clarify how a multiple simulation can be executed with the aid of this function.

Let us run the simulations using this system with the initial conditions **[0,1]** (initial velocity 0, initial displacement 1; see the assignment to the integrators in **denonit**) and the step size **step=0.01** over the time interval

[0, 5] seconds, for three different plate radii (3, 20, 60 cm). All the other parameters should remain fixed.

In accordance with the definition of the function **denonit**, we have to store the variation in the plate radius as a vector of length 3 and call the function accordingly.

This can be done in MATLAB with the following instruction:

```
>> r=[3.0, 20.0, 60.0];        % radius vector
>> rho=1.29*1000/1000000;      % rho is the same each time
>> Fc=155.2;                   % c is the same each time
```

The desired call is then

```
>> [t,Y] = denonit(r, rho, Fc, 5, 0.01, [0,1]);
```

Note that the corresponding Simulink system must be *open*. Otherwise, MATLAB reacts with an error message, which warns of erroneous Simulink objects.

The result can be visualized graphically using

```
>> plot(t,Y(:,1),'b-', t,Y(:,2),'k--', t,Y(:,3),'r-.')
```

Fig. 2.29 shows the result of this plot command.

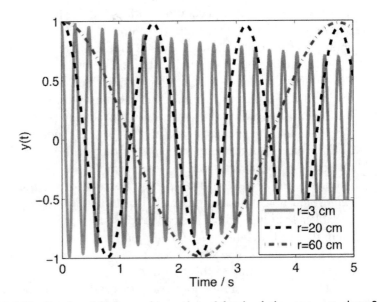

FIGURE 2.29   **Results of the iterated execution of the simulation system** s_deno3 **using** denonit **for three different plate radii.**

2.5.3 Transfer of Variables Through Global Variables

For the sake of completeness, here we allude to a last possibility for communication between MATLAB functions and Simulink.

It is possible to *declare parameters as global* by calling

```
global x y z p1 a2 ...
```

If this is included in a MATLAB function, then these variables can also be used upon transfer of the parameters to Simulink, if they will be used in the Simulink system.

This possibility is indeed very convenient but it conceals a number of dangers. Basically, in order to avoid undesirable side effects in all higher programming languages, including MATLAB, global variables should only be used in cases of compelling necessity. Under normal conditions their use should be urgently avoided, and the topic is not discussed further at this point for this reason.

PROBLEMS

Work through the following problems to practice the techniques by which Simulink interacts with MATLAB.

NOTE Solutions to all problems can be found in Chapter 4.

Problem 97
Call the Simulink system **s_Pendul** (in the accompanying software), which simulates a mathematical pendulum, from MATLAB with a pendulum length of 5 m and then with 8 m.

Problem 98

 Write a MATLAB function that can run simulations of the Simulink system **s_Pendul** for a vector of specified pendulum lengths.

Problem 99
Change the function **denonit** so that simulations with different spring constants can be run[9] as well as with different radii.

---

[9]Hint: use two nested **for**-loops.

Problem 100
Define a Simulink system which generates sine functions using the sine generator block **Sine Wave**. Then write a MATLAB function that can call this system repeatedly while allowing you to vary the frequency.

Problem 101
Design a Simulink system for solving the differential equation

$$\ddot{y}(t) + \dot{y}(t) + y(t) = f(t), \quad y(0) = 0, \dot{y}(0) = 0. \tag{2.17}$$

Here, $f(t)$ should be a step function (**Step** block) of height h.

Then write a MATLAB function with which you can call this system repeatedly while varying the parameter h.

Problem 102
Modify the function **denonit** so that in the Simulink system a procedure with variable step size can be used.

Since in this case the lengths of the vectors **y** are generally not all the same, the technique of collecting the results column-by-column in the matrix **Y** can no longer be used. Think about what other data structures might be used to solve this problem.

2.6   DEALING WITH CHARACTERISTIC CURVES

In many applications the functional relationship between parameters cannot be closed; that is, the parameters cannot be described in terms of a formula or functional prescription, but only in terms of measured values, which have been taken at discrete points.

The relationship then can be stored in the form of a lookup (indexed) table. A table of this sort is referred to as a *characteristic curve* or *family of characteristic curves*. Some well-known examples include the characteristics of a transistor, where, for example, the relationship between the collector-emitter voltage U_{CE} and the collector current I_C for given base currents I_B are given in the form of a series of characteristic curves or the operating characteristic of an engine, in which gasoline consumption is plotted as a function of the engine (rotational) speed and medium pressure.

In Simulink characteristic curves of this sort can be displayed using the **Lookup Table** and **Lookup Table (2-D)** blocks (see Fig. 2.30) from the block library **Lookup Tables**.

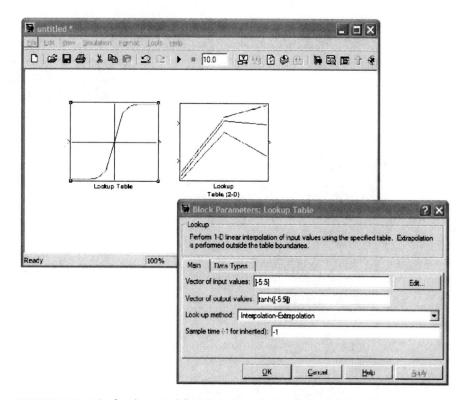

FIGURE 2.30 The lookup table **block from the** lookup tables **block set with the block parameter window for** lookup table.

It can be seen that defining a *characteristic curve* requires that two *equally long* vectors be specified, namely the vector of input values and the vector of output values that depend on it. The block then provides the output corresponding to an input value. If an input value is not included in the vector of input values, then (in the standard version) *linear interpolation* or *extrapolation* will be applied.

We now illustrate the use of lookup tables for handling characteristic curves with the aid of a simple example.

A solar cell has a highly nonlinear current-voltage (V-I) characteristic of which the following points are known:

U/V	0	2	4	6	8	9	10	11	12	12.3
−I/mA	562.5	537.5	512.5	487.5	462.5	450	437.5	400	275	0

For a better graphical representation we construct an (interpolated) characteristic curve with a step size of 0.1 V from these data using the MATLAB command `interplinterp1`:

```
>> volt1 = [  0     2     4    6     8    9   ...
             10    11   12  12.3 ];
>> mcurr1 = [562.5  537.5 512.5 487.5 462.5 450 ...
             437.5 400 275   0 ];
>> v1 = (0:0.1:12.3);
>> chary1 = interp1(volt1, mcurr1, v1, 'linear');
>> plot(v1, chary1)
>> grid
>> ylabel('-I / mA')
>> xlabel('U / V')
```

This intermediate step is not, in fact, necessary for using the **Lookup Table** block, since, as noted above, the interpolation of this block can be carried out automatically.

The calculated characteristic curve is shown in Fig. 2.31.

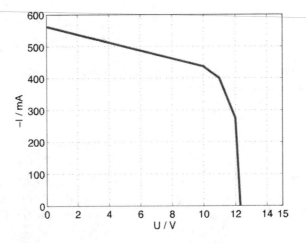

FIGURE 2.31 **Voltage-current characteristic of a solar cell.**

When the solar cell is attached to a load R Ω, a current $-I_R$ mA and voltage U_R V develop. These are given by the intersection of the characteristic curve and the (voltage-current) load line

$$I = \frac{1}{R}U \, . \tag{2.18}$$

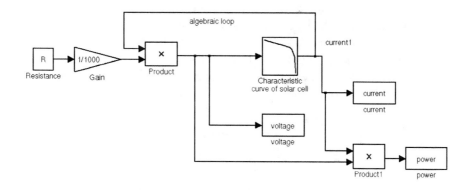

FIGURE 2.32   **The Simulink system** s_charc **with a** Lookup Table **for determining the operating point of a solar cell.**

This operating point can be calculated using Simulink. To do this we use the Simulink system shown in Fig. 2.32. The distinctive feature of this system is that it contains a so-called *algebraic loop*.

As can be seen in Fig. 2.32, the product $\frac{R}{1000} \cdot I$ is delivered to the input of the characteristic curves. The factor of 1/1000 has been introduced because the current I is in mA at the characteristic curve output.

On the other hand, because of the feedback this input depends *directly* on the output I of the characteristic curve. This type of feedback, without a time delay, is known as an algebraic loop. *In every iteration step* Simulink now tries to bring this algebraic loop "into balance." In our particular example, this means that in an internal iteration Simulink seeks the value of I such that:

$$U = \frac{R}{1000} \cdot I \quad I \text{ mA} , \tag{2.19}$$

$$I = f(U) \quad f \text{ is the characteristic curve} . \tag{2.20}$$

But this is precisely the desired intersection of the characteristic curve and the load line mentioned above.

For a simulation we only have to transfer the value of the resistance R Ω to the system. In the present example, $R = 18$ Ω. We do this by calling

```
R=18;
```

in the MATLAB command window.

We can then start a simulation of the system s_charc.

For a simulation we set the step sizes to 1 in the Simulink parameter window and the beginning and end times to 0. This might seem a bit odd at first, since with this setting *exactly one* simulation step will be executed. But, since the algebraic loop of interest to us will be solved in this single step, we only need this one step. A simulation then yields the following answer in the MATLAB command window:

```
>> Warning: Block diagram 's_charc' contains 1 algebraic loop(s).
Found algebraic loop containing block(s):
  's_charc/Characteristic curve of the solar cell'
  's_charc/Product' (algebraic variable)
```

This is a notice that an algebraic loop has been found in the system. The subsequent call of

```
>> [voltage, current]
```

then delivers the answer

```
ans =

    8.2653   459.1837
```

Thus, with a load of 18 Ω a current of 459.2 mA flows and a voltage of 8.26 V develops.

The power[10] is 3795.3 mW, as the following call shows:

```
>> power

power =

    3.7953e+003
```

Naturally, one might now get the idea of using the system to search for the value of the resistance corresponding to *power matching*, (i.e., for which the *maximum power can be extracted*).

As a final, and intimidating, example for this chapter, we show how this can be accomplished. The idea is to replace the constant R in the system s_charc with a function that runs through "all" R from 0 Ω to a certain limit, say 100 Ω. In each iteration step the intersection of the characteristic curve and the load line for this resistance, as well as the associated power, are calculated. We then only have to find the maximum of the vector power and the associated resistance. This, of course, we will again do with MATLAB.

---

[10]Note that U is in V and I is in mA.

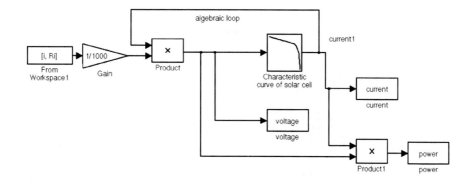

FIGURE 2.33 The Simulink system s_charc2 with a Lookup Table for determining the optimum operating point of a solar cell.

First, we modify the system **s_charc** to the system **s_charc2** shown in Fig. 2.33.

As can be seen in Fig. 2.33, the constant block has merely been replaced by a workspace source (**Sources** block library). This can read data in the matrix format **[timepoints, datapoints]** from the MATLAB workspace. In a simulation one of these data pairs will be processed in each simulation step.

In the present example, resistance values R_i for a certain set of indices i are read in. Thus, the index vector represents the "time" point vector and the resistance values the data point vector. For this we define the (column) vector pair **[i, Ri]** in the parameter window of the **From Workspace** block under **Parameters - Data** and, before the simulation, define the *column* vectors **i** and **Ri** in the MATLAB workspace as follows:

```
>> i=(1:1:200)';      % column vector of indexes
>> Ri=i*0.5;          % resistances in intervals of 0.5
                      % Ohms up to 100 Ohms
```

Next we start the simulation of **s_charc2**. The vectors **volt1** and **mcurr1**, which establish the characteristic curve must, of course, be defined in advance. As a step size we take 1 and as a (best) stop time **length(i)-1**.

The power can then be plotted as a function of the resistance using the following call:

```
>> plot(Ri, power)
>> grid
```

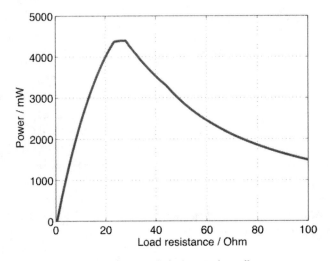

FIGURE 2.34 **Resistance-power characteristic for a solar cell.**

```
>> xlabel('Load resistance / Ohm')
>> ylabel('Power / mW')
```

Fig. 2.34 shows the result of this MATLAB command.

A peak power at a load of about 25 Ω can be read off the graph. This can be found more precisely with MATLAB, which gives

```
>> [pmax, indx] = max(power)

pmax =

  4.4010e+003

indx =

    54

>> Ri(indx)

ans =

    27
```

for a maximum[11] at 27 Ω and, in this case, a maximum extracted power of 4.4 W.

---

[11]The precision here is, of course, 0.5 Ω since that is our step size.

PROBLEMS

Work through the following problems to become familiar with characteristic curves in Simulink.

NOTE ➤ Solutions to all problems can be found in Chapter 4.

Problem 103

Consider the following characteristic curve:

$$k(x) = \begin{cases} 0 & \text{for} \quad x < 0, \\ \sqrt{x} & \text{for} \quad x \in [0, 4], \\ 2 & \text{for} \quad x > 4. \end{cases} \tag{2.21}$$

Develop a Simulink system that drives this characteristic curve with a sine wave signal and analyze the output signal. Experiment with various amplitudes of the sinusoidal input.

Problem 104

Consider the following "family of characteristic curves":

$$k(x, y) = x^2 + y^2 \quad \text{for } x, y \in [-1, 1]. \tag{2.22}$$

Using the **Lookup Table (2D)** block design a Simulink system, which reads out the values at the points (x, x) with $x \in [-1, 1]$ from the family of characteristic curves. In doing this, manipulate the family of characteristic curves with a suitably set **From Workspace** block.

Compare the considerations involved in setting the characteristic family block **Lookup Table (2D)** with the three-dimensional plots in Section 1.2.5.

Chapter 3

PROJECTS

3.1 HELLO WORLD

A traditional first programming assignment is the "Hello World" exercise.
Here, the goal is simply to display text. Typically, input and output are tricky
in a programming language, so once we have a quick way to show text to the
user, we can build it into something complex as needed. Also, it provides a
level of instant gratification to the programmer.

The **disp** function provides a simple way of displaying text to the user.

```
disp('Hello World')
```

Here is the response from MATLAB:

```
>> disp('Hello World')
Hello World
```

3.1.1 Personalized Hello World

We can personalize the greeting by adding a name. Suppose we have the
greeting ("Hello") stored as a string, and another string **name** which, as the
name implies, stores a person's name. We can concatenate the two strings by
putting the two variable names inside a set of square brackets:

```
>> greeting = 'Hello ';
>> name = 'Jane';
>> disp([greeting, name])
Hello Jane
```

Notice the space between the two words; it is actually part of the
greeting string. We could have put the space before "Jane," or even
inserted a space string between the two, as shown below.

```
>> greeting = 'Hello';
>> name = 'Jane';
>> disp([greeting, ' ', name])
Hello Jane
```

189

Using the square brackets is one option to concatenate strings, but MATLAB provides another way to do this with the **strcat** command.

```
>> greeting = 'Hello';
>> name = 'Jane';
>> disp(strcat(greeting, ' ', name))
HelloJane
```

Why did the two words run together? The answer lies in the documentation for **strcat**; it removes trailing spaces. Compare the two commands below.

```
>> disp(strcat('Hello ', 'Jane'))
HelloJane
>> disp(strcat('Hello', ' Jane'))
Hello Jane
```

Notice that the first command removed the space after "Hello," so that the words ran together. The second one has the space preceding "Jane," so it was not removed. We will examine the difference between **strcat** and the square brackets once more. This time we use the square brackets with two strings, with a space at the end of the first string, then try it again with a space at the beginning of the second string.

```
>> disp(['Hello ', 'Jane'])
Hello Jane
>> disp(['Hello', ' Jane'])
Hello Jane
```

We see that the square brackets performs the concatenation, without removing trailing spaces.

3.1.2 Hello World with Input

Now, let's make our "Hello World" program a bit more complicated. We will prompt the user to enter his/her name, then print a greeting based on the input. To accomplish this, we need the **input** command, one which allows the user to give us feedback. From the brief help available (using the command **help input**), we see that it will display a string for us before it gets the input. We really should tell the user what we want, otherwise our program will not be user friendly. Let's try it out, asking the user for his name. Notice the space after the question mark; we have this in place to make it look nice. MATLAB will not put a space between the prompt and what the user types unless we specify it like this.

```
>> input('What is your name? ')
What is your name? Steve
??? Error using ==> input
Undefined function or variable 'Steve'.

What is your name?
```

Pressing `control-C` returns us to the familiar MATLAB prompt. A quick look at the help entry for the `input` function reveals the problem. We need to specify that we expect a string, by passing the string `'s'` as a second parameter to the `input` function.

```
>> input('What is your name? ', 's')
What is your name? Steve

ans =

Steve
```

Next, we put the result of the `input` function into a variable. Then we can combine the name with a greeting, like we did previously. However, this time we will use the `sprintf` function, to specify how the name should be preceded and followed by text. The special characters `%s` are used to tell the `sprintf` function where to insert the string `name` into the greeting string. These characters do not appear in the final string; the `sprintf` function replaces them with the parameter string. Finally, we display the greeting.

```
>> name = input('What is your name? ', 's');
What is your name? Steve
>> greeting = sprintf('Hello %s, how are you?', name);
>> disp(greeting);
Hello Steve, how are you?
```

We can put several variables into a string like this by using `%s` several times in the first parameter to `sprintf`, then following it with a string for each place. This also works with other data, like characters, integers, and floating-point values.

3.2 SIMPLE MENU

Suppose that we want to interact with the user. Perhaps we want some information, a choice among several choices. MATLAB supports a `menu` function, which opens a window and displays a graphical menu to the user. The user then uses the mouse to click on one of the choices, and the **menu** function returns the number of the choice.

Below we have a simple example of the **menu** command. It has four parameters; the first gives a name to the menu, which could be a statement like "Please select an item." The other three items are the choices that will appear to the user. Since there are three of them in this example, there will be three buttons available to the user. When the user clicks on one of the buttons, the window will close, and the function returns the button number.

The window that pops up should look like the one shown in Fig. 3.1, though the window manager on your computer may give it a slightly different appearance.

FIGURE 3.1   An example graphical menu.

```
>> choice = menu('Lunch Special', 'fish', 'beef', 'eggplant')

choice =

     2
```

Here, we see that the second option, "beef," was selected by the user. Of course, we need to remember which number corresponds with which option. That is, whatever we want our program to do after the selection, we need to make sure we do the appropriate action.

But what if the user does not make a valid choice? The window appears with the standard buttons that all windows have: close, minimize, and maximize. A user could click on one of these. The minimize and maximize buttons do not get rid of the window, so we will not worry about these. The user could drag the window to a different spot, too. But the close button does get rid of the window; in this case, MATLAB returns a value of zero for the option.

It is possible that the menu will appear as text in the command window, if graphics are not available. Here is what happens when connected via secure shell ("ssh") to a remote computer that has MATLAB software. (This example will not work for you unless you have an account on the remote machine.) While it is possible to allow windows to appear from the remote computer, we do not do that here.

```
localHost:~> ssh remote.machine.gsu.edu
mweeks@remote.machine.gsu.edu's password:

mweeks@remote:~$ matlab
Warning: Unable to open display , MATLAB is starting without
  a display.
  You will not be able to display graphics on the screen.
Warning:
  MATLAB is starting without a display, using internal event
  queue.
  You will not be able to display graphics on the screen.


                       < M A T L A B >
              Copyright 1984-2006 The MathWorks, Inc.
                     Version 7.2.0.294 (R2006a)
                        January 27, 2006


    To get started, type one of these: helpwin, helpdesk, or
    demo.
    For product information, visit www.mathworks.com.

>> choice = menu('Lunch Special', 'fish', 'beef', 'eggplant')

----- Lunch Special -----

        1) fish
        2) beef
        3) eggplant

Select a menu number: 2

choice =

    2
```

```
>> exit
mweeks@remote:~$
```

As a side note, this could have been fixed by passing the -Y parameter to the ssh command, as seen below.

```
localHost:~> ssh -Y remote.machine.gsu.edu
```

This would have allowed X Window forwarding, so the window would have appeared as normal. Though generated on the remote machine, it would have been displayed on the local one. X Window software must be running on both machines, however. Also note that the ssh command is not part of MATLAB, though secure shell software does exist for all major computer platforms.

 We close this section with an example, located on the CD-ROM under the name "simplemenu.m." It presents a simple menu to the user, then processes the choice with the switch statement.

```
choice = menu('Lunch Special', 'fish', 'beef', 'eggplant');

switch choice
    case 1
       disp('You chose to have the fish.');
    case 2
       disp('You chose to have beef today.');
    case 3
       disp('You chose the eggplant dish.');
    otherwise % default
       disp('Looks like you closed the menu.');
end
```

You may notice that the first line is the same as what we have used before. The switch statement provides a regular way of dealing with several possibilities for the same variable. We could have used a series of if statements instead, but this way appears less cluttered.

The switch statement matches the variable choice to one of several values. We could have specified what to do with case 0, but we take care of that at the end, with the catch-all otherwise.

When you run this program, the menu pops up in a window. After you select one of the choices, the window disappears and one of the four messages appears in the command window according to which button you clicked.

3.3 FILE READING AND WRITING

Reading and writing files can be a bit tricky. We need to know what we are reading or writing, to know how to interpret the data. In other words, the computer does not automatically know what the data represents in a file.

The basic unit is the byte, an 8 bit value. Each bit can be either a 0 or 1, giving us 256 possible byte values. The character data type is closely related to the byte, though there are variations on the character type. We will use the `'uchar'`, or unsigned character type.

3.3.1 Writing a File

To test out the reading and writing capabilities, let's first write some data to a file. This will give us something to read, later. File operations require that we open the file when we want to start using it, and close it when we are finished. First, we open a file for writing.

```
filename = 'test.bin';
myfile = fopen(filename, 'w');
```

We have to tell the computer which file we want to use, as we do with the `filename` variable. We could do without that variable, and simply put the string as the first parameter of the `fopen` command. But this way makes the program more readable. The `fopen` command opens the file that we name, given the way that we want to open it. Here we simply want to write a file, so we use the `'w'` string to specify this. We could have several files open at the same time, so the computer assigns each a file handle, a number indicating to the computer which file we mean. The result of the `fopen` command will be a file handle, stored in the variable `myfile`. By the way, the file handle could be negative, indicating a problem. We should check the value, and quit if there is a problem.

```
if (myfile < 0)
    error(sprintf('Problem opening the file "%s".', filename));
    return
end
```

Above, we use the `return` statement to quit our program if we encounter an error. Such an error could be generated by a variety of things; we could be trying to write to a read-only medium, like a CD-ROM. Or we could be trying to write to a floppy disk that does not have any space left. Or perhaps the floppy disk was not inserted into the drive. Whatever the reason, we cannot resolve the problem within our program. It is a good idea to indicate the file name, since we could work with several files at once. It may even indicate the problem to the user, such as if the file name contains a sub-directory that we do not have permission to write.

Next, we will create some example data. We want to be sure that we can write and read every possible character. Since there are only 256 total, the following code presents a quick way to do so. We will actually write each value twice, to be sure that none of the characters cause a problem.

```
str = [0:255, 0:255];
```

We call the data `str`, short for string.

Now we can write it to the file. We have to indicate a few things: the file that we wish to write to, the data, and how the data should be written. Here we use the `'uchar'` data type, short for unsigned character, to write 8 bit values.

```
fwrite(myfile, str, 'uchar');
```

We could do more here, as desired. Since we wrote all data in that command, we will go ahead and close the file.

```
if (fclose(myfile) == -1)
    error(sprintf('Problem closing the file "%s".', filename));
end
```

Notice that the `fclose` command appears in an `if` statement. It will generate an error code if the file did not close correctly. If that happens, we should at least let the user know.

After we run the above code, how do we know if it worked? One thing that we can do is check the directory contents for a new file. If we see a file called "`test.bin`", with a size of 512 bytes, then we can be fairly confident that everything worked. Of course, a better way to know is to read the contents back, and verify that we have what we expect. We will do this next.

3.3.2 Reading a File

Our steps to read the file will be very similar to those we used to write the file. We will open the file, read data, then close the file when done. First, we open it. See if you can spot the difference between this and the code used to open a file for writing.

```
filename = 'test.bin';
myfile = fopen(filename, 'r');
```

The first line is exactly the same, so we read the same file that we created above. Did you see the slight difference in the second line? We open this file with `'r'` in single quotes, indicating that we want to read it. This small detail carries much importance; if we had used `'w'` again, the computer would have started making a new file with the same name. This would effectively destroy the old file.

As before, we next check to make sure the **fopen** command worked. It could fail for a variety of reasons, such as if we got the file's name wrong. Or perhaps if there is no disk in the drive.

```
if (myfile == -1)
    error(sprintf('Problem opening the file "%s".', filename));
    return
end
```

Assuming that everything worked, we will continue on with the read.

MATLAB actually provides a few different commands for reading and writing. Since we used **fwrite** before, we will use **fread** for a sense of symmetry.

```
[mybuffer, chars_read] = fread(myfile, 512, 'uchar');
```

The values passed to the **fread** command include the file handle, the number of values to read, and the data type. We again use the **uchar** data type to read 8 bit values. The command returns both the data to us, as well as the number of values that were read. You may expect the number of values read to always be equal to the number requested, but we could request more data than the file contains. Also, we should note that commands to read or write data in a file do so sequentially. That is, the computer keeps track of how much data we have read, or how much we have written. When we read or write more, the computer uses that information to get or put the data in the correct spot. It does not matter to the computer if we read all 512 bytes at once, or read 256 then the next 256, or even if we were to read them one value at a time.

We could have used the **fscanf** command instead, as below. The `'%c'` indicates that the computer should read and return the data as characters.

```
[mybuffer, chars_read] = fscanf(myfile, '%c', 512);
```

We would not want to use *both* here, though, since we only have 512 bytes in this particular file.

Finally, we close the file. Compare this code to that used to close the file that we wrote. You should see that the lines are identical.

```
if (fclose(myfile) == -1)
    error(sprintf('Problem closing the file "%s".', filename));
end
```

Again, we check for error and let the user know if the file close operation encountered a problem.

Now we examine the data, and verify that we were able to write and read all values successfully. Notice how we created the data in variable str in the file writing code, but used a different variable name, mybuffer, in the file reading code. We will run both programs, then compute the error between these two variables.

```
>> write_test
>> read_test
>> error = sum(abs(str - mybuffer))
??? Error using ==> minus
Matrix dimensions must agree.
```

What happened? MATLAB helpfully tells us that the dimensions do not agree. Let's examine the dimensions of these variables.

```
>> size(str)

ans =

     1   512

>> size(mybuffer)

ans =

   512     1
```

Well, that explains the problem. Variable str has 1 row and 512 columns, while variable mybuffer has 512 rows and 1 column. The fread command returned mybuffer in this way. We can still compare the two, by using the .' operation, which performs the transpose.

```
>> error = sum(abs(str - mybuffer.'))

error =

     0
```

When we use the transpose as above, we align the two variables by changing `mybuffer` to have 1 row and 512 columns. We find the difference between each value using subtraction. Then we perform the absolute value on the result, and sum it all together. For example, if there were subtraction results of 1 followed by −1, they would add to 0 and make us think that there were no differences. Thus, we use `abs` so that errors cannot cancel each other out.

With the sum of all differences between `str` and `mybuffer.'` adding to zero, we conclude that the two variables contain exactly the same data. If you are so inclined, you can examine the contents of both variables after running the code.

 If you want to experiment with the code mentioned in this section, we provide it on the CD-ROM as the programs `write_test.m` and `read_test.m`, respectively.

3.4   SORTING

MATLAB provides some sorting functions, but there are times when we may need our own. Suppose that we have a list of numbers, and we want to sort them in ascending order. But what if we need to remember their original indices? This could happen when the indices are used for other data, like the key from a database table. Perhaps the index corresponds to a name in another set of data.

Consider the following data.

```
x = [23, 95, 61, 49, 89, 76, 46, 12];
```

We could use the built-in `sort` function, but this is of limited value to our application.

```
>> disp(sort(x))
    12    23    46    49    61    76    89    95
```

If we want to link the numbers returned from the sort back to our indices, we have to match them by looking through the original data. Instead, we will combine the numbers to sort with the indices, as two rows of the same matrix. We could put the indices in the top row; this does not really matter at this point.

```
x = [23, 95, 61, 49, 89, 76, 46, 12;
      1,  2,  3,  4,  5,  6,  7,  8];
```

Now to sort the `x` values, we define a value `n`, then see how the `x(n)` value compares to the value next to it. In the following instance, we compare it to the value to its left.

```
n = 2;
if (x(1, n-1) > x(1, n))
    % swap x(n) and x(n-1)
    temp = x(:, n-1);
    x(:, n-1) = x(:, n);
    x(:, n) = temp;
end
```

Notice that we selected a value of 2 for n, since a value of 1 would lead to a problem. MATLAB does not allow 0 as an index as would be generated by x(n-1, 1).

The commands to swap two values of x are fairly standard. First, we store one of the values of x in a temporary variable. Next, we set that value to the other x value, then we set the latter value to the temporary variable. The reason for the temporary variable should be evident; without it, we would overwrite (and lose) one of the values in the second step.

Would this code above make the switch in this instance? Think about it for a moment; we see if x(1,1) is less than x(1,2), then make the swap if so. But in our example values, x(1,1) has the value 23, while x(1,2) has the value 95. These two values are already in the correct order, so we would not make the switch.

Now let's expand the role of n to encompass all elements of x. We will let it vary from 2 to whatever the length of x happens to be. We do not start it at 1 for a good reason; as mentioned above, the **if** statement uses the value n-1 to index x, so starting n at 1 would result in an error.

```
for n = 2:length(x)
    if (x(1, n-1) > x(1, n))
        % swap x(n) and x(n-1)
        temp = x(:, n-1);
        x(:, n-1) = x(:, n);
        x(:, n) = temp;
    end
end
```

When we run the above code, we will swap several values of x. It brings us a step closer to the solution, but we are not done yet. To see why, we only need to examine x.

```
>> x

x =
    23    61    49    89    76    46    12    95
     1     3     4     5     6     7     8     2
```

We see that the value 95 has "bubbled up" to the right-most position in our list. This sorting technique gets the name "bubble sort" from this phenomenon. We can imagine why the code moves 95 all the way to the right; whenever we compare a value to the maximum value of the list, we will move that maximum value to the right. On the next loop iteration, the maximum value will appear in its new location, and again be moved to the right.

The bubble sort does not sort very efficiently, and there are many faster algorithms. But the bubble sort has an advantage of being easy to understand. We can improve this sort by noticing that the maximum value of the list always ends up at the right after one pass of this code. Now if we consider the x values, except for the right-most one, as a new list, then the next highest number will be moved to the right end. We do not need to compare it to the maximum value from the previous run, since we already know they are correctly ordered. For example, after a single pass of the code above, the 95 occupies the right-most spot. So we run the code again, not considering that spot, and see that the 89 moves to the second right-most spot. We can continue this process until we have completely sorted the list.

To complete this example, we need to modify the code above to stop earlier than the length of x, except for the first pass. Also, we need another loop around the above code so that we repeat it until we are done. The code below combines both of these ideas. We set a new variable called x_end to range from the length of our data x, down to the value 2.

```
for x_end = length(x):-1:2
    for n = 2:x_end
        if (x(1, n-1) > x(1, n))
            % swap x(n) and x(n-1)
            temp = x(:, n-1);
            x(:, n-1) = x(:, n);
            x(:, n) = temp;
        end
    end
end
```

We put the code above in a function, called sort_with_index.m. When we run the above code, with the original x, we get the following result.

```
>> x = [23, 95, 61, 49, 89, 76, 46, 12;
         1,  2,  3,  4,  5,  6,  7,  8];
>> sort_with_index(x)
```

```
ans =

    12    23    46    49    61    76    89    95
     8     1     7     4     3     6     5     2
```

It looks like everything in row 1 was sorted correctly. Remember that row 2 contains the original indices, so we expect that they will be in a strange order.

Before we conclude that this code works, let's try a worst case example.

```
>> x2 = [99, 88, 77, 66, 55, 44, 33, 22;
          1,  2,  3,  4,  5,  6,  7,  8];
>> sort_with_index(x2)

ans =

    22    33    44    55    66    77    88    99
     8     7     6     5     4     3     2     1
```

Here we have the data (row 1) already sorted, only in reverse order. This means that every comparison results in a swap. We see that the function results in all values switched around. Since it works for the worst case example, we can conclude that it achieves its goal.

3.5   WORKING WITH BIOLOGICAL IMAGES

This project comes from biology research. We have an image, in this case one of Caenorhabditis elegans (C. elegans) worms, taken with a video microscope. The video microscope takes 30 frames per second, at 10-times magnification. Each frame is saved as a ".tiff" file. Ultimately, we want to track the movement of the worms from frame to frame, but for the moment we concentrate on just one image at a time.

Neurons within the C. elegans worms are made fluorescent, so that we can readily see where they are. The fluorescent areas of interest show up as white circles. These circles typically are close together, like peas in a pod. But when several groups appear on the same image, the software treats them as one group, leading to poor information. Refer to Fig. 3.2, that shows an example of how these worms may appear. For clarity, we show an idealized, reversed image. A human observer quickly sees two worms on the figure, but having the computer make this distinction requires a complex algorithm. At this stage of research, we simply want to select a sub-image when more than one worm appears in the frame.

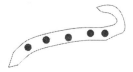

FIGURE 3.2   An idealized example image of C. elegans worms.

3.5.1   Creating a Sub-image

First, we set up some variables to define the file names and type of image. A finalized version will likely make these parameters. We should change these lines according to the image we use.

```
file_to_read = 'worms.tif';
file_to_save_as = 'worms_subim.tif';
file_type = 'TIFF';
```

We could guess what file name to save the file as, such as automatically appending _subim before the file extension. Also, we can usually figure out the file's type by looking at the extension. But we will use these three lines to get started; we can always refine the program later.

Next, we read the image data and show it to the screen.

```
[image_data] = imread(file_to_read, file_type);
figure(1);
imshow(image_data);
```

The first command above loads the file (image data) with the imread command. Then we create figure number 1 if it does not yet exist. MATLAB will create the figure automatically if we do not issue this command, but it also indicates that we want to use figure number 1. We may run this program, with multiple figures, more than once. Finally, we display the image with the imshow command.

Once we show the image, we let the user select two corner points to define a sub-image. We need to let the user know that we expect input, so we use the `title` command to convey this information. After that, the `ginput` command puts a cross-hair over the image, and returns the coordinates of wherever the mouse happens to be when the user clicks the button.

```
title('Select the upper left corner of the area of interest');
[x1, y1] = ginput(1);
title('Select the lower right corner of the area of interest');
[x2, y2] = ginput(1);
title(file_to_read);
```

Notice that we repeat the commands, to get two corner points from the user. We keep track of them as (`x1`, `y1`) and (`x2`, `y2`), respectively. The final `title` command changes the text on the image to the file name. We change that text to let the user know that we do not expect more input, at least not now.

Now that we know the dimensions of the area of interest, we use the corner points to define the sub-image. We use the `round` function since the coordinates may not be whole numbers, and we need them to be integers to work as indices. The code below handles this.

```
x1 = round(x1);
y1 = round(y1);
x2 = round(x2);
y2 = round(y2);
[MAXROW, MAXCOL, DEPTH] = size(image_data);
sub_image = image_data(y1:y2, x1:x2, 1:DEPTH);
figure(2); imshow(sub_image);
```

Here, we also use the `imshow` command to let the user see the sub-image that he selected. The `size` function gives us the lengths of each dimension of `image_data`, which we need to know now to determine the "depth" of the image. Images typically have a depth of 1 or 3, depending on whether it is grayscale or color.

Next, we plan to store this sub-image. We use this opportunity to show a menu, like the earlier project. Create the question to pose with the `sprintf` command, then show it as a menu with "yes" and "no" as the two options.

```
question = sprintf('Write sub-image (fig 2) to file "%s" ?', ...
    file_to_save_as);
option = menu(question, 'Yes', 'No');
```

This way, the user can choose not to store the sub-image, in case a point was accidentally selected, or if the sub-image just does not look right. Of course, we need to act upon the user's choice. Below, we check the response. If the user clicked on the "Yes" button, then we use the `imwrite` command to store the sub-image data. Also, we print an informative message to the screen. Otherwise, we print a message indicating that we did not write the image.

```
if (option == 1)
    imwrite(sub_image, file_to_save_as, file_type);
    disp(sprintf('Yes - sub-image written to file "%s".',...
        file_to_save_as));
else
    disp('No - the sub image was not written.');
end
```

Did you notice the default? That is, suppose the user does not choose either menu option, but instead clicks the "close window" icon on the menu. In that case, the result for variable **option** will be 0. The code interprets all values of **option**, besides 1, to be "No," so the sub-image will not be saved.

After finishing the program, we need to test it out. Watching people try this program, it was evident that some do not read the directions! Rather than fault the user, we can make our program more tolerant. The user will select two points representing a rectangle, and we should accept these two points as long as they are on the image. First, let's perform a check to make sure that the points are on the image. We will do this now for just one corner coordinate, though we need to repeat the checks for all of them.

```
[MAXROW, MAXCOL, DEPTH] = size(image_data);
if ((x1 <= 0) || (x1 > MAXCOL))
    warning('corner point x1 is not on the image.');
    disp('Using default.');
    if (x1 <= 0)
        x1 = 1;
    else
        x1 = MAXCOL;
    end
end
```

After we find the size of the image (in the first line above), we see if our **x1** coordinate is off of the image. This could happen if the user positions the mouse on the figure's border. We print a **warning** message in that case, and pick a default value closest to the coordinate. For example, when the user clicks the mouse button to the left of the image, the code above responds as if the first column were chosen. You may notice that the first line in this code segment was already included in previous code. We put it here now for clarity. In the final program, we only need to include it once.

Next, we should check the points to make sure they are in the right order, and re-arrange them as needed. There are four corners in a rectangle, and our user could select any two, in any order. But if the user selects two points on the same horizontal or vertical line, then we have a rectangle with zero size. We will catch this problem later. For now we can ignore this possibility and focus on two corner points diagonally across from each other. Either the diagonal line between corner points goes from the lower left to upper right, or it goes from the upper left to the lower right. The user could select either corner point first, resulting in four possibilities. Let's examine some specifics.

Suppose that the user clicks points (10, 15) and (2, 4), corresponding to an upper left to lower right diagonal. Remember that the `ginput` command returns (x, y) coordinates. They do specify column and row of the pixel value under the cursor, when rounded. Therefore, the desired sub-matrix has rows from 4 to 15 and columns from 2 to 10. It does not matter what order the user chooses these points.

Now suppose the user clicks on the points (1, 40) then (35, 8). This corresponds to a diagonal from the lower left to upper right. The image showing function (`imshow`) has the point (1,1) in the upper left corner of the image, and that the row coordinate increases as we go down along the left, while the column coordinate increases as we go across to the right. Thus, the sub-matrix of the image, for the above points, would be from (1:35, 8:40). Notice that we want the same sub-matrix, even if the user clicks point (35, 8) first.

Fig. 3.3 shows the four possibilities for corner points. We see that we always define our rectangle from the minimum x (and y) values to the maximum x (and y) values, regardless of which points are selected or what order. We see from these examples that we always form a rectangle with the ranges from the lowest to highest value, regardless of which point the user selects first, or whether the diagonal rises to the left or to the right.

The simple programming solution checks to make sure that x1 is smaller than x2, and switches them when they are not. Then we repeat the same check for y1 and y2 values.

```
if (x2 < x1)
    temp = x1;
    x1 = x2;
    x2 = temp;
end
```

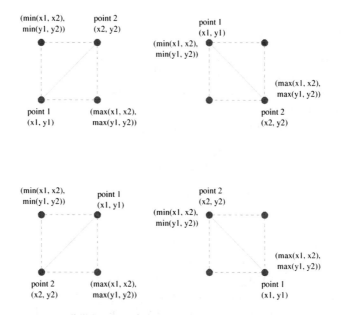

FIGURE 3.3   **Four possibilities for selecting two corner points.**

Notice the **temp** variable, used to momentarily remember **x1**'s value. Your first impulse may be to write **x1 = x2** followed by **x2 = x1**, but this means that the computer overwrites **x1**'s value with **x2**, and both variables end up with the same value.

What if, after all the checks we perform, the sub-image has no size? The code below handles this.

```
[rows, cols, depth] = size(sub_image);
if (˜((rows>=1) && (cols>=1) && (depth>=1)))
    disp('Sorry - image has no size!');
    return
end
```

Actually, we should not need this check. Even if the user clicks on the exact same point twice, the resulting sub-image should be at least one pixel in each dimension. In this case, we let the user write an image of a single pixel if he wants.

Now the program should be ready to use. We get corner points from the user, and round these values to correspond to rows and columns in the image data. We check to see that the chosen points are on the image, and give default values when they are not. When the corner points are given in a different order than what we asked, we sort the coordinates to make them usable. We also check to make sure the selected sub-image has a non-zero size. When we have a valid sub-image after all of this, then we allow the user to save the sub-image to a new file.

Now that we have a working program, we can change it to a function. To do this, we add a **function** line to the top of the program that specifies the variables for inputs and outputs.

```
function sub_image = ...
    copy2subimage(file_to_read, file_to_save_as, file_type)
```

This has the effect of making our program an abstract tool; the inputs for it are clearly specified, as are the outputs. Any variables that it creates will be cleared from the user's workspace when the function finishes. Also, any variable with the same name as one in the function will be preserved.

We have one more detail to address: the output variable should have a default value, in case the function returns early. This is good programming practice, so we set variable **sub_image** to a null matrix at the beginning. Under normal circumstances, it will not be returned to the user.

```
sub_image = [];
```

 The program that encompasses all of these points is **copy2subimage.m**, available on the CD-ROM.

3.5.2 Working with Multiple Images

Recall what are we trying to accomplish with this. We have a set of images with worms, where the DNA was altered to activate a fluorescent gene. Biologists want to connect the fluorescent neurons, to see how the worm progresses. As the worms move, they can be tracked from frame to frame. This program allows a person to select a sub-image for one frame, thereby isolating a worm for further processing. The frames come to us as a set of images. With the **copy2subimage** function available, we can call it repeatedly using a different file name each time. The code below demonstrates this idea.

```
mysub = copy2subimage('file1.tif', file1_sub.tif', 'TIFF');
mysub = copy2subimage('file2.tif', file2_sub.tif', 'TIFF');
mysub = copy2subimage('file3.tif', file3_sub.tif', 'TIFF');
...
```

This direct approach leaves much to be desired. What if we have a set of, say, 100 images? We would have a lot of typing to call the function using each image. Instead, we should automate the process by having the computer generate a set of file names.

Here, we generate ten file names. We break up the file name into three parts, the first part that remains the same for each file name, that we call **root**, followed by the number, followed by the file extension in a variable named **ext**. We put these together with the **sprintf** command, then display the results. Of course, we could call another function with the generated variable **file_name**, such as the **copy2subimage** function.

```
root = 'file';
ext = '.tif';
for num = 1:10
    file_name = sprintf('%s%d%s', root, num, ext);
    disp(file_name);
end
```

When we run the above code, we get the following output.

```
file1.tif
file2.tif
file3.tif
file4.tif
file5.tif
file6.tif
file7.tif
file8.tif
file9.tif
file10.tif
```

This confirms that we generate successive file names, based on the number. Now, what if we want to generate a second file name, to hold the sub-image data? We can alter the line that assigns variable **file_name** to a string, such as the following.

```
>> file_name  = sprintf('%s%d_subimage%s', root, num, ext)

file_name =

file10_subimage.tif
```

Notice that we added text to the format string. We could have also created another string variable to go between **num** and **ext**, or we could have simply altered string **ext** by prepending characters.

3.6   WORKING WITH A SOUND FILE

People who work with audio use frequency plots to analyze sound. In this project, we will look at sound data in terms of frequencies.

 First, we will load an example sound file (available on the accompanying CD-ROM). MATLAB provides the `wavread` function to work with sound files stored in the "WAVE" format (ending with a ".wav" extension).

```
[x, fs, b] = wavread('violin.wav');
```

The `wavread` function returns the sound data (in the two-dimensional array `x`), followed by the sampling frequency and the number of bits. We need the sampling frequency (`fs`) to know how fast to play the sound back. The sound data typically has two channels, one for each stereo speaker, which is why `x` is two-dimensional.

Once we have the sound data, we can listen to it.

```
sound(x, fs, b);
```

This sound data was created by recording a violin.[1] An instrument like this causes strings to vibrate, which in turn causes variations in air pressure. We perceive the air pressure changes as audible sounds. When we talk about music, we often speak of the pitch or frequency of notes. But we can see the frequency information, too, by plotting the spectrum. The theory behind this process is outside the scope of this book, but we will use the fast Fourier Transform (FFT) to convert the time domain data to the frequency domain.

To see the magnitudes of the frequencies within sound data, we only need to perform a few steps. First, we obtain the sound data, such as reading it from a file. Then we use the `fft` function to convert the data to frequency information. We will plot it, but only the first half of the data since the second half would simply be a mirror image. So we set up an array, called `half`, that ranges from 1 to half the data length. When we plot the data, we have to convert it from complex numbers to magnitudes. MATLAB provides the absolute value function `abs` which will do this, and we see it in the code segment below with the `plot` function.

```
[x, fs, b] = wavread('violin.wav');
X = fft(x);
half = 1:round(length(X)/2);
```

---

[1]Thanks to Laurel Haislip for playing the violin for this sound sample.

```
plot(half*fs/length(X), abs(X(half)), 'k');
axis tight
```

The final command above, **axis tight**, makes the plot look a bit better. Try the code above without that line, then add it to see the difference. When we run this code, we get the spectrum as seen in Fig. 3.4.

 The CD-ROM includes a copy of this code in the file **make_music_freqs.m**.

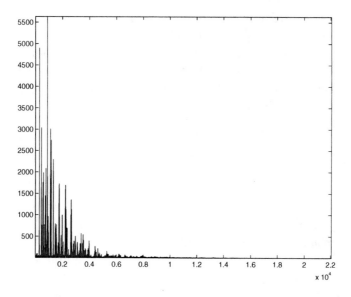

FIGURE 3.4   **Frequency magnitudes for the entire violin recording.**

However, viewing the frequency data for the whole signal has limited value. We might be able to say that a particular note was played, but we have no idea *when* it occurred. A nice improvement would be to view the frequency information a little bit at a time. To accomplish this, we will break the sound data into blocks, then convert it to frequency information and display it.

First, we choose a suitable block size. We create the variable **blocksize**, and give it a value of 8820. By inspecting the values returned by the **wavread** command above, we know that the sampling rate **fs** equals 44100 samples per second. Our block size 8820 is exactly one fifth of that, so we will examine five blocks per second. You could use a larger or smaller value, depending on your preference.

We define our range next. We need to know where it starts and where it ends. We store this information in **start_index** and **end_index**, respectively. For the moment, we use values of 1 and **blocksize**, though these values will be updated as we progress through the data. Finally, we define **range** based on these index variables.

```
start_index = 1;
end_index = blocksize;
range = start_index:end_index;
```

Later, we will put the definition for **range** within a loop, so that the program updates the range every loop iteration.

Based on the range, we will find the frequency information. This means that we perform the **fft** on a subset of **x**, which contains the sound data. Actually, **x** contains two channels' worth of sound data, but we will ignore the second channel.

```
X = fft(x(range,1));
```

The command above finds the frequency information for one block of **x**'s data. You may notice that we use the same letter for both the time domain data and the frequency domain data. MATLAB supports this common convention by distinguishing between upper- and lowercase variable names.

With frequency information, we get two pieces of data: a magnitude and a phase angle. The magnitude indicates the relative strength of the particular frequency, while the phase angle gives timing information that we need to correctly reconstruct the signal. MATLAB provides the function **angle** to return to phase angle from a complex number. If this is unfamiliar to you, consider that a complex number specifies a real and an imaginary coordinate, much like the x and y coordinates on a two-dimensional plane. Just as we can convert the Cartesian (x and y) coordinates to polar coordinates, we can do the same with a complex number. Polar coordinates specify a distance from the origin and a rotation angle. If a person stands at some starting point, facing east, and takes 3 steps forward then turns left and takes 4 steps to the north, he arrives at the same spot as if he were to turn 53.1 degrees counter clockwise and take 5 steps forward. Fig. 3.5 shows this idea, and the code below verifies this example.

```
>> p = 3+4j;
>> disp(abs(p))
    5
```

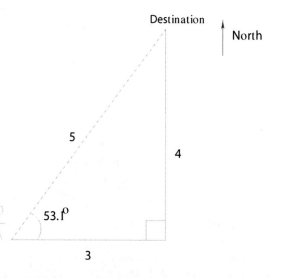

FIGURE 3.5 **Walking 3 units east and 4 units north reaches the same destination as turning 53.1 degrees and walking 5 units.**

```
>> disp(angle(p) * 360 / (2*pi))
   53.1301
```

We see that the complex value $3 + 4j$ has a magnitude of 5. Finding the angle of that complex value, then converting it from radians results in 53.1 degrees.

We are likely to have non-zero data at all frequencies. The magnitudes clearly show when a frequency has a tiny (or negligible) contribution to the signal. But when we plot the phase angles, we do not know how significant the corresponding frequency is. We would see the phase angles even for frequencies that have almost zero magnitude. A better way to view this data is to check the magnitude first, and zero-out the corresponding phase angles when the magnitude falls below some threshold that we set. The following line sets this up, then the line after it plots the phase angles.

```
zeros_ones = (abs(X(half)) >= 10);
plot(half*fs/length(half), angle(X(half)).*zeros_ones, 'k');
```

Variable **zeros_ones** contains an array of zeros and ones, logical values. We could compare its elements to **true** or **false**, but we instead use it as a point-by-point multiplier. Each phase angle will be multiplied by a zero or a one, effectively suppressing the phase angles when the corresponding magnitudes are less than our threshold of 10.

To show how this works, let's look at a specific pair of examples.

```
>> disp((5 >= 10))
     0

>> disp((12 >= 10))
      1
```

In the first example, we display the result from the comparison of 5 greater than or equal to 10, and see that it is zero (false). The second example show that comparing 12 greater than or equal to 10 returns one (true). Now we use these results with slightly more involved examples.

```
>> disp((5 >= 10) * 123)
     0

>> disp((12 >= 10) * 123)
    123
```

The first comparison returns a zero, and multiplying it by 123 still gives us zero. The second comparison, since it returns a one, gives us 123 after the multiplication. The variable **zeros_ones** uses this idea for the entire list of magnitudes. Like the number 123 in the above examples, the phase angles either appear as zero, or whatever value they have.

Fig. 3.6 shows a typical spectral plot, with the magnitudes on top, and the phase angles on the bottom. You can see how most of the phase angles are shown as zero, so that the ones with significant corresponding magnitudes stand out.

After we plot both the magnitudes and phase angles, we invoke the **pause** command to give the system a chance to respond. Our program will halt for a very small amount of time (0.001 seconds), which allows other programs to run. Though the figure will have changed, the computer may not show the change on the screen. With the pause, the computer will update the screen.

```
pause(0.001);
```

Now the program should go on to the next block's worth of data. To do this, we advance the start and ending indices by the block size as in the following code.

```
start_index = start_index + blocksize;
end_index = end_index + blocksize;
```

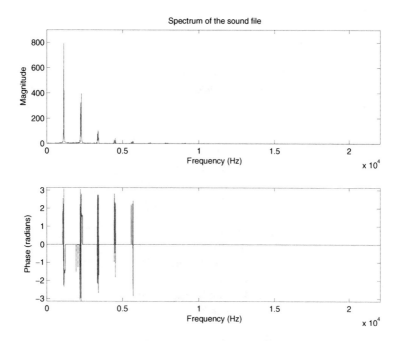

FIGURE 3.6   A typical spectral plot of the violin recording.

We could update `start_index` with a different value than `blocksize`, such as with `blocksize` divided by 2, creating a sliding-window effect.

 Now that we have the components, we can put them together in the code below. The CD-ROM contains this program, saved as `music_spectrum.m`.

```
%
% Show the frequencies present in a sound clip,
% for a bit at a time.
%

blocksize = 8820;
% Read the sound file
[x, fs, b] = wavread('violin.wav');

% Initialize the start and end indices.
start_index = 1;
end_index = blocksize;
```

```matlab
% Split this signal into sections,
% and show each section.
while (start_index < length(x))
    % Do not go beyond the end of the signal.
    if (end_index > length(x))
        end_index = length(x);
    end

    range = start_index:end_index;
    half = 1:ceil(length(range)/2);
    % Find the fast Fourier transform of x,
    % for 1 channel only.
    X = fft(x(range,1));
    % Create a array of 0's and 1's,
    % based on the magnitude of X.
    % We use it below to multiply the phase
    % angles, so that frequencies with tiny
    % amplitudes do not show a non-zero phase angle.
    zeros_ones = (abs(X(half)) >= 10);

    % Show the Spectrum
    % First the Magnitudes
    subplot(2,1,1);
    plot(half*fs/length(half), abs(X(half)), 'k');
    axis([0 fs 0 900]);
    title('Spectrum of the sound file');
    xlabel('Frequency (Hz)');
    ylabel('Magnitude');
    % Now the Phase angles
    subplot(2,1,2);
    plot(half*fs/length(half), angle(X(half)).*zeros_ones, 'k');
    axis([0 fs -pi pi]);
    xlabel('Frequency (Hz)');
    ylabel('Phase (radians)');

    % sound(x(range), fs);
    % include a pause so that figure is updated
    pause(0.001);

    % Advance the start and end indices.
    start_index = start_index + blocksize;
    end_index = end_index + blocksize;
end
```

3.7   PERMUTATIONS

When we can simplify a possible solution down to a list of numbers, then manipulating the list allows us to check out other possible solutions.

For example, consider the traveling salesman problem, where we want to travel to each of a list of cities, but do not care what order. If we number the cities to visit from 1 to N, then a possible solution would be $\{1, 2, 3, \ldots, N - 1, N\}$. This solution would mean we would travel to city 1, then city 2, then city 3, etc. It may be that this just happens to be the best (shortest or cheapest) solution. We could re-arrange these numbers, called a *permutation*, to try other possibilities. But allowing numbers to repeat, as they might in a *combination*, would not work for this problem. In other words, for $N = 3$, we would want to consider $\{1, 2, 3\}$, $\{1, 3, 2\}$, or $\{2, 1, 3\}$, etc. But we would not consider $\{1, 2, 2\}$ or $\{3, 3, 3\}$, since these would mean that we would miss one or more cities.

Let's design a function to generate permutations. This problem is more difficult than it first appears, since the number of permutations quickly increases as N increases. For example, if we have two numbers, then permutations are $\{1, 2\}$ and $\{2, 1\}$. When $N = 3$, permutations are $\{1, 2, 3\}$, $\{1, 3, 2\}$, $\{2, 1, 3\}$, $\{2, 3, 1\}$, $\{3, 1, 2\}$, and $\{3, 2, 1\}$, a total of 6. For $N = 4$, we have 24 permutations. For $N = 5$, we have 120 permutations. Obviously, this growth is not linear. The number of permutations is $N!$.

Tackling a problem like this can be intimidating. First, we need to have a good idea of exactly what we are trying to accomplish. Writing out the ideas as comments helps the process. A great way to approach it is to make it work for a small case, then make it larger. We can start with a single number, $\{1\}$. Then, when we go to two numbers, we have $\{1, 2\}$ and $\{2, 1\}$. Effectively, we take the former list and add the new number before and after it. When we go to three numbers, the pattern becomes clear. We want to take each row and insert the new number before, after, and in between each value. Every time we do this, we generate a new row.

We will make a matrix manually with the first two values. Then, we will isolate the first row of the matrix. If we can successfully insert the new value in all positions for the first row, then we can repeat this for the second, third, and other rows.

```
>> matrix = [1, 2; 2, 1];
>> matrix_row_1 = matrix(1, :)

matrix_row_1 =

     1     2
```

Now we will define the new number to add, as well as variables to keep track of the number of rows and columns.

```
>> value = 3;
>> [maxrow, maxcol] = size(matrix);
```

Now we can insert the new value into the row that we have. The first and last positions will be handled separately.

```
>> disp([value, matrix_row_1]);
     3    1    2

>> disp([matrix_row_1(1), value, matrix_row_1(2:maxcol)])
     1    3    2

>> disp([matrix_row_1(1:2), value, matrix_row_1(3:maxcol)]);
     1    2    3

>> disp([matrix_row_1, value]);
     1    2    3
```

Note that the `matrix_row_1(3:maxcol)` generates an empty matrix, since `maxcol` is only 2. This is why the third line above gave us the same result as the last line. We will not actually use the third line for the case $N = 3$, but it does show the pattern that we are after. Also, we would repeat the above for each row of the original. Below, we generalize the insertion of the new value into the given row.

```
>> k = 1;
>> while (k+1 <= maxcol)
        disp([matrix_row_1(1:k), value, matrix_row_1(k+1:maxcol)]);
        k = k + 1;
   end
     1    3    2
```

Here, it only generated one new row, but only because there are a small number of columns. Now, we put this together with the code generating the first and last lines, and we get the result below.

```
>> disp([value, matrix_row_1]);
     3    1    2
```

```
>> k = 1;
>> while (k+1 <= maxcol)
       disp([matrix_row_1(1:k), value, matrix_row_1(k+1:maxcol)]);
       k = k + 1;
   end
     1     3     2

>> disp([matrix_row_1, value]);
     1     2     3
```

We see that we have the three new rows when given the row {1, 2} and value 3. To finish this example, we need to repeat this for every other row. In this case, that just means the second row, but we will keep the code general.

```
% initial conditions
matrix = [1, 2; 2, 1];
value = 3;

[maxrow, maxcol] = size(matrix);
for row = 1: maxrow
    % Isolate the row.
    matrix_row = matrix(row, :);

    % Give the first permutation with the new value.
    disp([value, matrix_row]);

    % Give the permutations where the value is
    % inserted between columns.
    k = 1;
    while (k+1 <= maxcol)
        disp([matrix_row(1:k), value, matrix_row(k+1:maxcol)]);
        k = k + 1;
    end

    % Give the last permutation.
    disp([matrix_row, value]);
end
```

When we run this code, we get the following output.

```
     3     1     2

     1     3     2
```

```
   1      2      3

   3      2      1

   2      3      1

   2      1      3
```

We can verify that all six permutations appear in the output. To finalize the program, we will get rid of the **disp** statements, and instead store the new permutations in a matrix we will call **possible**. The syntax is to set **possible(n, :)** equal to the new array. For example, the first new row would be as the code shows below.

```
>> possible(1, :) = [value, matrix_row];
```

We will call the variable used for the row number **possible_row**. We have to be careful how we use it; we do not want to overwrite a row that we already stored. Therefore, we will increment this variable before using it. Below, we have the code listing, saved as the function **perm.m**.

```
%
% perm.m
%
% Given an matrix like [1, 2], [2, 1], and a value 3,
% return a matrix of all possible permutations, i.e.
% [1, 2, 3], [1, 3, 2], [2, 1, 3], [2, 3, 1],
% [3, 1, 2], [3, 2, 1];
%
% usage:
%         possible = perm(value, matrix);
%
% e.g.    mypossible = perm(3, [1, 2; 2, 1])

function possible = perm(value, matrix)

% find the dimensions of the matrix
[maxrow, maxcol] = size(matrix);

% possible_row is the last row in the new matrix.
% We will always increment it before using it with
%    possible(possible_row, ...
possible_row = 0;
for row = 1: maxrow
```

```
% Isolate the first row.
matrix_row = matrix(row, :);

% Give the first permutation with the new value.
possible_row = possible_row + 1;
possible(possible_row, :) = [value, matrix_row];

% Give the permutations where the value is
% inserted between columns.
k = 1;
while (k+1 <= maxcol)
    possible_row = possible_row + 1;
    possible(possible_row, :) = [matrix_row(1:k), value, ...
                                matrix_row(k+1:maxcol)];
    k = k + 1;
end

% Give the last permutation.
possible_row = possible_row + 1;
possible(possible_row, :) = [matrix_row, value];
end
```

Now let's test the new function.

```
>>  mypossible = perm(3, [1, 2; 2, 1])

mypossible =

     3     1     2
     1     3     2
     1     2     3
     3     2     1
     2     3     1
     2     1     3
```

Of course, we can (and should) verify that this code works for larger values. In the code below, we verify that the number of rows returned matches our expectations. It also demonstrates how we might use this function to generate all permutations for a given N.

```
>> mypossible = perm(4, mypossible);
>> size(mypossible)
```

```
ans =

    24     4

>> mypossible = perm(5, mypossible);
>> size(mypossible)

ans =

   120     5
```

Let's formalize that last idea. Suppose we want to generate all possible solutions for the traveling salesman problem, which can be expressed as all permutations of the numbers 1 through 6. The code below utilizes our **perm** function, to generate a matrix **M** that contains each permutation.

```
n = 6;    % total values
M = [1]; % initialize our permutation matrix
% Add a new value, one at a time.
for k=2:n
    M = perm(k, M);
end
```

By running the code above, we see that the permutation matrix **M** ends up with 720 rows, like we expect.

3.8   APPROACHING A PROBLEM AND USING HEURISTICS

As with many programming problems, there is more than one way to solve this problem. For example, we might start with a row of all the values $\{1, 2, 3, \ldots, N\}$, and generate new rows by switching values, e.g., $\{2, 1, 3, \ldots, N\}$. We face similar difficulties as we did above, in that we have to echo the switches throughout the rest of the row, as well as throughout each new row. But we could make this work.

Another option is to ignore some of the possible solutions by implementing rules that should focus on the best possibilities. We call such rules *heuristics*. For example, suppose that the solution corresponding to $\{2, 1, 3, 4, \ldots, N\}$ is worse than any solution that begins with $\{1, 2, \ldots\}$. Perhaps we can ignore all permutations that start with $\{2, 1, \ldots\}$ as a result. While this could give us a "good enough" solution, we will not know for sure if it is the best one. It will just be the best one that we found. The point is that there are other ways of thinking about a problem, which generate different solutions.

3.9   MAKING PERMUTATIONS FASTER

We saw how to create a matrix of permutations. The solution works well for small numbers. But as the number of permutations gets larger, the function becomes much slower. For eight columns, the function takes about five minutes to run. For nine columns, it takes over seven and a half hours! Clearly, the function could be better.

There are two issues here. One is that it may be possible to find a faster way to compute the permutations matrix. The second issue is the nature of the problem itself; as the size grows, the time grows enormously.

3.9.1   A Faster Way

To understand how we can make the function faster, we have to know a bit more about how the language works. MATLAB allows us to use *dynamic allocation*, where it reserves memory for us, as needed. We do not have to be concerned about how large our matrices are, or even know how large they will become. If we tell MATLAB that a matrix has a previously undefined row, it will make room for it. The problem is that it may have to move things around in memory, as it dedicates more memory to the matrix.

Consider a white board as an analogy. Imagine that you stand in front of a white board, and that it has many things already written on it. Someone asks you to write down what he says. You pick a clean spot on the board, and start writing. But you have no idea how much you will write; you picked the spot based on a guess that it would be large enough. But what if you run out of room? In that case, you might move to a larger spot, copy what you have already written, then erase the words from the smaller spot. This could happen again and again. (Perhaps if you cannot find a larger spot, you might recopy other things on the board and erase the originals to make space.) Obviously, you would have picked a spot that was "large enough" if you knew in advance how much you were going to write.

A computer handles dynamic memory allocation in a similar way. It picks a chunk of available memory and uses it. But if it needs more, it will have to find a larger chunk and recopy the data. It may have to do this repeatedly. The free memory will be used by other programs, too (imagine if other people were using the white board at the same time). Also, memory is a limited resource. Thus, the program must request a fixed amount. With virtual memory, the computer acts as if it has more memory (RAM) than it really does. It will write sections of memory to the hard drive, until those sections are needed again. While this greatly increases flexibility, it also drags down performance, since accessing the hard drive takes much longer than accessing RAM.

The computer may even move things around to make room, a process called *garbage collection*. This colorful name comes from the idea of collecting memory that programs have used and disposed of, and it shows the attitude of the not-so-distant past. The term "recycling" better describes the process. The garbage collection process can happen at any (unpredictable) time, which impacts performance.

To summarize, dynamic memory allocation provides a convenience to the programmer, but at a potential performance hit. Our permutation function asks for more and more memory as it runs. We can overcome this issue by figuring out how much memory we will need up front, and requesting it.

The `perm.m` function works by building on a smaller permutation matrix. Even if the smaller matrix only has one value, we still use it to make the new matrix. The new matrix will have one additional column, so we can easily find the new matrix's width. To find the height, we have to know a bit more about our problem. In this case, the number of rows returned will be the factorial of the number of columns.

We can find the size of the current matrix, which we call `matrix`, with the following command.

```
[numRows, numCols] = size(matrix);
```

To compute the output matrix's columns, we can use `numCols+1`. Finding the output matrix's rows requires us to find the factorial, e.g., `factorial(numCols+1)`. These two commands specify the dimensions of our output matrix. All that we need to do now is reserve the memory, which we can accomplish by using the `zeros` function. This function creates a matrix of zeros, matching the number of rows and columns given to it as parameters. The line below assigns this matrix of zeros to our output variable `possible`.

```
possible = zeros(factorial(numCols+1), numCols+1);
```

All we need to do is add these two lines to our `perm` function! This new function is shown below. We will call it `perm1.m`, until we are satisfied that it works as expected.

```
%
% perm1.m
```

```
%
% Given an matrix like [1, 2], [2, 1], and a value 3,
% return a matrix of all possible permutations, i.e.
% [1, 2, 3], [1, 3, 2], [2, 1, 3], [2, 3, 1],
% [3, 1, 2], [3, 2, 1]
%
% usage:
%        possible = perm1(value, matrix);
%
% e.g.   mypossible = perm1(3, [1, 2; 2, 1])

function possible = perm1(value, matrix)

% Allocate memory for the matrix
[numRows, numCols] = size(matrix);
possible = zeros(factorial(numCols+1), numCols+1);

% find the dimensions of the matrix
[maxrow, maxcol] = size(matrix);

% possible_row is the last row in the new matrix.
% We will always increment it before using it with
%    possible(possible_row, ...
possible_row = 0;
for row = 1: maxrow
    % Isolate the first row.
    matrix_row = matrix(row, :);

    % Give the first permutation with the new value.
    possible_row = possible_row + 1;
    possible(possible_row, :) = [value, matrix_row];

    % Give the permutations where the value
    % is inserted between columns.
    k = 1;
    while (k+1 <= maxcol)
        possible_row = possible_row + 1;
        possible(possible_row, :) = [matrix_row(1:k), value, ...
                                     matrix_row(k+1:maxcol)];
        k = k + 1;
    end

    % Give the last permutation.
```

```
        possible_row = possible_row + 1;
        possible(possible_row, :) = [matrix_row, value];
    end
```

Now we should test this out, to see the improvement.

By running the original (**perm**) and improved (**perm1**) functions, for varying numbers of **N**, we can record the time needed by each version. The code below tests the original **perm** function. To test **perm1**, all that we need to do is add a **1** after the two occurrences of **perm**. We use the **cputime** function here, which will be explained shortly.

```
N = 8;                              % total columns desired
start_time = cputime;               % get start time
subrouteMatrix = perm(2, [1]);      % get initial matrix
for num=3:N
    % add a new column
    subrouteMatrix = perm(num, subrouteMatrix);
end
elapsed_time = cputime - start_time;  % find difference in times
```

Table 3.1 summarizes the results. We should ignore the results before six columns, since the run times are so small. Examining columns seven through nine shows that the new version of **perm** provides a drastic improvement: over three seconds compared to about an eighth of a second; over five minutes compared to less than one second; and well over seven hours to about six and a half seconds. Clearly, adding the two lines that allocate memory has a great benefit.

3.9.2 Measuring Time

Notice that we used the difference between **cputime** readings instead of **tic** and **toc**. This gives us a more accurate view of the elapsed time, since it measures the time the CPU spends working on MATLAB, instead of the

TABLE 3.1 Run times (seconds) for the original and improved "perm" functions.

	2	3	4	5	6	7	8	9
original (**perm**)	0.0100	0.0200	0	0	0.1000	3.2600	309.8900	27214
improved (**perm1**)	0	0	0	0	0.0500	0.1300	0.7300	6.4900

actual (clock) time. The actual time includes time spent by the computer on other tasks.

To show this, we can re-run the original **perm** function and track the elapsed time in three different ways.

```
start_time = cputime;              % get start time
t1 = clock;
tic
```

After this, we call the **perm** function in a **for** loop, to build a permutation matrix with 8 columns. Next, we measure the times again, as the code below.

```
elapsed_time = cputime - start_time;  % find difference in times
t2 = clock;
toc
```

To get the time difference between our two clock readings, **t1** and **t2**, we use the **etime** function: **etime(t2, t1)**. Running this experiment on a G4 based Apple iBook (laptop), other programs were used at the same time to tax the CPU, including Firefox, DashBoard, Quicktime, and iTunes. In other words, the CPU had to simultaneously provide services for web browsing, a chess game, showing a movie clip, and playing music, all while running the MATLAB code. It does this by spending a little bit of time with each task, then switching to the next one. The results are 491.7 seconds for the elapsed time with **tic** and **toc**, 363.1 seconds for the **cputime** measurement, and 491.7 seconds for the difference in the clock readings. As expected, the time taken by these other tasks prevented MATLAB from having the CPU's full attention. While over 8 minutes went by, the CPU only spent about 6 minutes on the **perm** function.

You may have noticed that the run time in the above experiment differs from the time seen in Table 3.1. For Table 3.2, the data were collected from two different, though similar, machines. Both are Apple computers with G4 PowerPC processors, one a desktop and the other a laptop model. The desktop computer was also used for Table 3.1. Since the machines are not identical, this is one explanation. Another possibility is that the readings are extremes; every time the experiment runs, the run time is different. This can result from the system not being in the exact same state. For example, it could be that more memory is available during one of the runs, so the operating system responds to the request more quickly.

Running the same program repeatedly on two different computers shows us what results are typical. Table 3.2 shows the run times for the original program on two different computers. The average time for the laptop computer

TABLE 3.2   Run times (seconds) for the original "perm" function, on two different computers.

| | \multicolumn{10}{c}{Run number} | Average |
	1	2	3	4	5	6	7	8	9	10	Average
Laptop	363	425	426	347	351	350	350	351	349	350	366
Desktop	318	317	305	302	306	307	306	305	306	307	308

is 366 seconds, while the desktop computer averages 308 seconds for the same task. We can safely conclude that the laptop computer is slower for this task than the desktop one.

3.9.3 The Growth of the Problem

This problem grows in complexity at a very rapid rate. For every N columns, we generate $N!$ rows. The number of operations is key; just looking at the number of operations for inserting the new number, there is one for each row. In other words, running the program for $N = 4$ results in $4! = 24$ operations, but this jumps to $5! = 120$ operations for $N = 5$. When $N = 10$, there are more than 3.5 million operations. When $N = 15$, we have over a trillion operations. When $N = 100$, the number of calculations has 157 digits.

Let's suppose that we have a 10 GHz computer. How long would it take to calculate all of the permutations, for $N = 100$? We can make a few simplifying assumptions, such as that every operation takes exactly one clock tick, and that the computer has nothing else to do but run our program. Also, we will assume that memory is not an issue, even though realistically this would be a problem for even much smaller values of N.

Below, we have code that will simulate this for us.

```
%
% Suppose we look at the perm.m function for N=100.
% Since it takes N! operations, how long is this?
%

N = 100;

ops = factorial(N);      % number of operations
```

```
ten_GHz = 10000000000;    % Say we have a 10GHz processor
seconds = ops / ten_GHz;  % time in seconds
minutes = seconds / 60;   % time in minutes
hours = minutes / 60;     % time in hours
days = hours / 24;        % time in days
years = days / 365;       % time in years
```

The number of years for the permutation program to run is 3×10^{140}. Surprised? Actually, even a "low" value like $N = 20$ would take over seven and a half years to run. Running the program for another "low" value like $N = 27$ takes longer than our universe is known to have existed, using an estimate of 20 billion years. Knowledge of such problems helps us avoid trying to solve them with a direct "brute force" method. Instead, we would turn to a trick to reduce the problem complexity: a heuristic.

3.10   SEARCH A FILE

This project demonstrates reading a file. We will open a file, search for a string, and display lines that match. If you are familiar with Unix, you probably are reminded of the handy **grep** utility. In fact, if you are using MATLAB on a Unix-based computer, you can execute a shell command with the exclamation point. The following code shows the results of using the **grep** utility, under MATLAB, on an Apple Macintosh (using OS X).

```
>> !grep zeros perm1.m
possible = zeros(factorial(numCols+1), numCols+1);
>> !grep zeros perm.m
>>
```

The first parameter, **zeros** specifies the text to find. The second parameter says that we want to search the file **perm1.m**. As you can see, the system responds with every line from the **perm1.m** program that contains the text **zeros**. In this case, it only appears once. When we use the same command a second time on file **perm.m**, we do not find any occurrences. We will develop a similar utility, but in MATLAB code.

3.10.1   A Side Note About System Commands

With the exclamation point preceding a system command, MATLAB will execute the command. We saw above how that would work with a built-in command like **grep** under Unix machines. But the system command could be one of our own commands. Below, we have an example of a simple Unix shell script, called "`testme.sh`." The first line declares that it uses the `sh` shell. The other line says to output some text to the screen, just like how a `disp('Hi there.')` command would work.

```
#!/bin/sh

echo "Hi there."
```

We can view the file contents under MATLAB with the `type` command, as demonstrated below. (This works on all platforms.)

```
>> type testme.sh

#!/bin/sh

echo "Hi there."
```

Now we can execute this shell script, from MATLAB. Note that we need a Unix OS for this example to work.

```
>> !testme.sh
Hi there.
```

As you can see, MATLAB ran another program and told us the results. Often, programmers use shell scripts to get data, or transform data from one file to another. It would be possible to invoke a shell script with MATLAB, as above, then have MATLAB read and process any resulting files. For example, we could have a simple shell script to **grep** all occurrences of a certain file name in a web-page access log, and write the results to another file. Then we could open and read that file with MATLAB, to count and plot the number of web-page requests per hour.

3.10.2 DNA Matching

To motivate this project, consider working with DNA data. GenBank® contains genetic sequence data available to the public. This genetic information consists of a sequence of the bases A, C, G and T, that comprise genes. We will use a sequence of random data, shown in Fig. 3.7, though real data can be obtained from the Internet. This data appears in the file "`DNA_example.txt`."

When dealing with such data, we should be able to have the computer scan for sequences of interest. For example, suppose that we want to know how many times, and where, the sequence **CAT** appears?

```
TTCCCAAGGCCCTGATGAGGAACAATAAGGCCATTCGAGC
GTGTTATAGGTCCCGAATGAAACCATACCATCGCCGAGGA
TTGCATACACTTCTCGGGGGGGGGGTTCGACTGAGCTCACT
TGGTAAGACATGATTTTCCGGTTCTCAGGGTTGTTGGTCG
AAAAACTCATAGGGCACCGTCGAACGCTACGTATTATTTT
AGTCGGATTGCGGCTACACCTTCAAGGTGCAGTTTCCGAA
AGAAGCCGTAAAATAGCAGGTCGTGTGCGATATGAACTTA
TGAGCAAAATATCGTAACACATTAGACAACAATCGAAATT
ATGTCCCACATAACGTGACAACGATTGTAACGTGGTTGGT
TCTGCGCTCAGGAGATAAGACCAACTTAGTGTGGCTCAGT
ACACTCCCGGTTCGAGCTTCTGAACAACTCCCTAGCCCTT
TCTAACCCTCCGGACCTTTGGGAACGCTCAGGAATCAAGG
AGCACATGTGTCTAGGCCAGTAAATATGTAACCCCCTGTA
GCCAGCTTCCCTCCTCTACTGAGAGTCGGTAGTCGGCCTA
ACGTCACTCCTCAGCCTTGTTCAGTAACAGTCGACCCCAT
```

FIGURE 3.7   **Example DNA data.**

3.10.3 Our Search Through a File

Our goal is to look through a file, and find all occurrences of a string. To do this, we will use the following algorithm. We specify it in *pseudo-code*, a set of steps that read almost like commands for a computer, but meant to convey the idea to a human.

```
1. get text and file name from the user
2. open the file, stop if there is an error
3. read characters into a buffer
4. does the buffer match our search text?
   if so, remember the location
5. read more character(s) into the buffer
6. if the read was successful, goto 4
7. close the file
```

First, let's make a simple draft of the program we want. We can simply specify the search text and file name with string assignments. Next, reading the file and checking for error can be accomplished with the **fopen** command, followed by **error** and **return** if that command fails. To read characters from the file, we can either use the **fscanf** or **fread** commands. We will use **fscanf**, though it is possible to make it work with either command. Once we have some file data, we can compare it to our search text. MATLAB allows a simple == comparison for strings or matrices, assuming that they are the same size. To read more characters, we use the same function that we used in step 3. The number of characters read will be returned, and we can check to see if the number read matches the number we requested. If not, we know we have reached the end of the file. Finally, we will close the file with the **fclose** command.

```
% matgrep1.m
% Scan a file for pattern matching
% First attempt - just scan a file for text
%

% 1. get text and filename from the user
function [locations] = matgrep1(pattern, filename)

% We can show what we are searching for
disp(sprintf('pattern = "%s"', pattern));
disp(sprintf('filename  = "%s"', filename));

% initialize variables
locations = [];
total_chars_read = 0;
num = 1;                  % index for results

% 2. open the file, stop if there is an error
myfile = fopen(filename, 'r');
if (myfile < 0)
    error(sprintf('Sorry, file "%s" does not exist.', filename));
    return
end


% 3. read characters into a buffer
[mybuffer, chars_read] = fscanf(myfile, '%c', length(pattern));
total_chars_read = total_chars_read + chars_read;
% disp(sprintf('read %d chars from file.', chars_read));

while (chars_read > 0)

    % 4. does the buffer match our search text?
    %    if so, remember the location
    if (mybuffer == pattern)
        % total_chars_read points to the last char in
        % the string. We want to remember where the
        % string starts instead.
```

```
        locations(num) = total_chars_read - length(pattern) + 1;
        num = num + 1;
        % disp('Found the pattern');
        % disp(mybuffer);
    end

    % 5. read more character(s) into the buffer
    [new_char, chars_read] = fscanf(myfile, '%c', 1);
    mybuffer = [ mybuffer(2:length(mybuffer)), new_char ];
    total_chars_read = total_chars_read + chars_read;

    % 6. if the read was successful, goto 4
end


% 7. close the file
status = fclose(myfile);
if (status ~= 0)
    error('Could not properly close file.');
end
```

Let's test this function on the DNA sequence file.

```
>> a = matgrep1('CAT', 'DNA_example.txt');
pattern = "CAT"
filename = "DNA_example.txt"
>> a

a =

    32    64    69    84   129   168   300   329   485   598
```

We see that the pattern is located at several locations in the file. The last one (598) shows that the function returns the very last match as well. To double-check our results, we can load the example file into a text editor and have it tell us where it finds the pattern. Doing so reveals that the numbers above match the pattern's positions within the file.

It is always a good idea to test our functions for a variety of data, to ensure robustness in them. We will try the function again, but on a special file called "**test.bin**," a binary file with every possible byte value listed twice. Some byte values cause problems during file read operations. For example, the byte value corresponding to the decimal number 26 represents an end-of-file marker to some systems. Other systems might interpret a byte value of zero in a similar fashion. If we can read each of the characters of this file, we should be able to successfully read any file.

```
>> b = matgrep1('abc', 'test.bin')
pattern = "abc"
filename  = "test.bin"

b =

    98    354
```

We see that our function finds the pattern **abc** twice, showing that our function reads everything as desired.

3.10.4 Buffering Our Data

You might wonder why we do not load the whole file, and search through the data in memory. This certainly could be done with the examples we have here. But there may be times when this becomes impractical, such as when the file's size exceeds the computer's memory. This question brings up an interesting point; a computer reads its memory much more quickly than it reads files stored on a disk. In fact, we have to be careful how we read disk files for good performance. On some systems, reading a byte at a time severely reduces performance unless we buffer the data.

Consider the locality of data. If we access an array value at position 1, we are likely to need the value at position 2, then later the value at position 3. We call this spatial locality. If accessing these values takes a lot of time and effort, it makes sense to read more than we need for the moment. For a computer system, this means getting the disk up to speed (if it is not already), positioning the read/write head over the correct track, waiting for the correct sector to pass under the read/write head, reading the data, and passing it along to its destination. Reading, say, 512 bytes from a file does not significantly add to the time needed to read a single byte. Fortunately, the operating system takes care of these disk operation steps for us, but it may not automatically buffer the data.

Can we improve our function's performance by reading more data instead of requesting a byte at a time? We will find out! This will involve re-writing our search function. You may wonder if the effort required is worth it. After all, it is better to have an inefficient program that takes a few minutes to run rather than spending hours making it efficient, especially if we do not plan to use it over and over again. We will assume that this program gets enough use that the time modifying it will be time well spent.

Among the differences between the two programs, the new one needs additional variables to keep track of the locations within the buffer. First, we will compare our search string to a small part of the buffer, since the buffer will be much larger by comparison. We need to know where in the buffer we are, so we make a `start` index. Think of scanning a page for a phrase; you might run your finger along the text as you scan it with your eyes, to keep your place. The `start` index serves this purpose. We will also use a `stop` index, essentially `start` plus the length of search string. We also need a variable for the position of the find with relation to the file. That is, suppose we read 100 characters every time, and locate a match starting at the fifth character of the buffer. Would that be the fifth character of the file, or the 105th, or 205th? We use the variable `position` to keep track of this.

For the new program we must consider how we know when to stop. Instead of stopping when the file read function returns no new data, we will make it a bit more complex with a flag (`done`) to indicate this. Before, we would read from the file during every loop iteration. Now we will only read the file as needed, and update our flag when we do. In other words, we give the flag a `false` value, and set it to `true` when we try to read more data but none is returned. It signals when we are done, and we process it when we are ready.

In the original version, we just compared the buffer to the search string. Now that the buffer is much larger than the search string, we have to compare it with a sub-part of the buffer. But there is a potential problem in that we need to make sure the buffer has enough data left. Suppose we have a buffer of 100 characters, and a search string of 10. We might be tempted to use the comparison below.

```
if (mybuffer(start : start+length(pattern)-1) == pattern)
```

In fact, we do use something like this, but we have to be careful. For example, when `start` has a value of 92, then we want to compare the search string to the buffer's contents in the range 92:101. As MATLAB will be quick to point out, the "index exceeds matrix dimensions." To avoid this problem, we must make sure that the `stop` value does not exceed the buffer's size. When it does exceed the buffer, we need to read more from the file. This can be tricky, since we must keep the original data that we have not yet examined, while moving it over to give space for new data. You will notice that algorithm step 5 becomes more complex.

Below we present the updated version, that reads a buffer at a time. The reader may want to compare algorithm steps 1, 4, and 5 to the original version of the code.

```matlab
% matgrep2.m
% Scan a file for pattern matching
% Second draft - read a block of data,
%      and scan a file for text
%

% 1. get text and filename from the user
function [locations] = matgrep2(pattern, filename)

% We can show what we are searching for
disp(sprintf('pattern = "%s"', pattern));
disp(sprintf('filename = "%s"', filename));

% initialize variables
locations = [];
total_chars_read = 0;
num = 1;                 % index for results
position = 1;
% position is the location of the start character, in the file.
% This is similar to start, below, except that
% start will be reset when we read from file.
start = 1;              % Where to start the comparison
% Where to stop the comparison
stop = start+length(pattern)-1;

buffer_size = 512;      % How much to read at once
% We assume that the pattern will be much smaller
% than the buffer_size. But in case it is not, we can
% try to make it work.
if (length(pattern) > buffer_size/4)
    buffer_size = length(pattern) * 10;
end

% 2. open the file, stop if there is an error
myfile = fopen(filename, 'r');
if (myfile < 0)
    error(sprintf('Sorry, file "%s" does not exist.', filename));
    return
end



% 3. read characters into a buffer
[mybuffer, chars_read] = fscanf(myfile, '%c', buffer_size);
```

```
total_chars_read = total_chars_read + chars_read;
% disp(sprintf('read %d chars from file.', chars_read));

% Set condition to end the following while loop
if (chars_read > 0)
    % We want to go through the loop
    % to search for the pattern.
    done = false;
else
    % If nothing was read, then there is no data
    % to search through.
    done = true;
end

% Search through the data for our pattern.
% Read more from the file, as needed.
% Quit when we completely run out of data.
while (~done)

    % 4. does the buffer match our search text?
    %    if so, remember the location
    % Check to see if we will over-run a boundary
    if (stop < length(mybuffer))
        % The part of mybuffer that we need is within bounds.
        if (mybuffer(start:stop) == pattern)
            % total_chars_read points to the last char in
            % the string. We want to remember where the
            % string starts instead.
            locations(num) = position;
            num = num + 1;
            % disp(sprintf('%d Found the pattern', num-1));
            % disp(mybuffer);
        end
    else
        % We need to read more
        % 5. read more character(s) into the buffer
        % We want to have buffer_size bytes in the buffer,
        % but we need to preserve the part of buffer that
        % we did not scan.
        [new_chars, chars_read] = fscanf(myfile, '%c', ...
            buffer_size - length(pattern));
        mybuffer = [ mybuffer(start:length(mybuffer)), new_chars ];
        total_chars_read = total_chars_read + chars_read;
```

```
        % disp(sprintf('Read %d bytes', chars_read));
        if (chars_read == 0)
            done = true;
        end
        start = 0; % We will add one below
        % Adjust position, so that we don't count this iteration.
        % We did not make a comparison this time through.
        position = position - 1;
    end


    % 6. if the read was successful, goto 4
    % Increment our starting point
    start = start + 1;
    stop = start+length(pattern)-1;
    % Increment the position to remember
    position = position + 1;
end

% 7. close the file
status = fclose(myfile);
if (status ~= 0)
    error('Could not properly close file.');
end
```

Now that we have a (hopefully) improved file search function, we need to test it out. How fast do these two functions work? MATLAB provides **tic** and **toc** functions, named like the sounds of a clock, to find elapsed time. Function **tic** starts the virtual stopwatch, and **toc** ends it and reports the time difference. Assuming that the computer does little else while the program runs, this will work for our purposes.

For the example data that we have been using, the functions run very quickly. To make things interesting, we try the search functions on a much larger file. The "**DNA_big.txt**" file is 1000 times the size of the "**DNA_example.txt**" file. In the lines below, note that the lines beginning with the words **pattern** and **filename** are outputs from our program. The lines beginning with the word **Elapsed** come from the **toc** function. So if you are trying this yourself, you only type the text after the MATLAB prompts ("»") to the end of that line.

```
>> tic; a = matgrep1('CATTGA', 'DNA_big.txt'); toc
pattern = "CATTGA"
```

```
filename  = "DNA_big.txt"
Elapsed time is 45.633932 seconds.
>> tic; b = matgrep2('CATTGA', 'DNA_big.txt'); toc
pattern = "CATTGA"
filename  = "DNA_big.txt"
Elapsed time is 22.784429 seconds.
```

As we see from the two time measurements, the buffered version runs in about half the time as the original! But before we celebrate, are we sure that both functions work?

With the original version, we could verify that it found text within a file reliably by loading the test file into an editor, and scanning it ourselves. Below, we will simply compare the results from the original and buffered versions.

```
>> sum(abs(a-b))

ans =

     0
```

Had the results (**a** and **b**) been different, we would have had a non-zero sum above. We can rest assured that both functions are correct.

Running the functions on the large data file, for different search strings, produces similar results. Table 3.3 shows a few different runs with the original and buffered functions. You may notice that it actually takes longer to find shorter strings within the file. These run times can vary, depending on other factors such as what other software the computer runs simultaneously. But the shorter search strings mean that it will be found significantly more times.

When it comes to speed, there are other factors to consider. For example, changing the buffer size will have an effect on the run-time. Of course, the computer that you use will likely have different run times than what we saw in Table 3.3. It is possible that hardware improvements, such as a disk cache, could greatly improve the run time. Also, be aware that other searching algorithms do exist, with possibly a better performance.

3.10.5 A Further Check

Before we conclude that our program works, we should check for correctness at the transition between two consecutive reads. Since we have our own buffer for data read from the file, there will come a time when we need to add data to our buffer without disturbing the data that is already there.

TABLE 3.3   Run time comparisons (seconds).

search pattern	index	
	original (matgrep1)	buffered (matgrep2)
CGA	49.5	26.0
GTT	49.0	26.1
CATTGA	45.6	22.8
GTACGT	45.5	22.6
GTACGTTT	45.7	22.7
CGAGTTAA	44.9	22.7

Suppose that our buffer is 10 characters long, and our search string has six letters. We first compare the search string to the first six characters of the buffer. Then we compare it to the next six characters, starting at position 2. Next, we compare it to the six characters including and after position 3, then the six starting at position 4, then the six starting at position 5. But after this, we have a problem; we want to compare it to the six characters starting at position 6, but there are only five characters in the buffer between 6 and 10: buffer(6), buffer(7), buffer(8), buffer(9), and buffer(10). We do not have data at buffer(11), since position 11 is beyond the buffer limits. We need to read more data from the file, but we want to preserve the data in buffer(6:10). We can get rid of buffer(1:5), though, so we will move buffer(6:10) to buffer(1:5), then read data from the file and store it at buffer positions 6 through 10. This makes for some tedious programming; it would be easy to have a bug. Therefore, we will check to make sure the transition happens smoothly.

First, load the **matgrep2.m** program into the MATLAB editor, and highlight the **fscanf** line after algorithm step 5, then select the **Debug** drop-down menu, then choose **Set/Clear Breakpoint**. For clarity, the line appears below.

```
[new_chars, chars_read] = fscanf(myfile, '%c', ...
```

When selected for a breakpoint, the MATLAB editor will show a red dot on the left, next to the line number. This line precedes the tricky part where we move the buffer's contents around, so we can check it line by line. Now we will call the function from the command window.

```
>> a = matgrep2('CAT', 'DNA_example.txt');
pattern = "CAT"
filename  = "DNA_example.txt"
K>>
```

MATLAB stops execution of the function, and gives us the debug prompt. First, we will check out a few variables.

```
K>> start

start =

   510
```

We see that the value for **start** is almost as large as the buffer size. Since we use that value as an index into the buffer, it means that we have examined most of the buffer's contents. Even though our search string (**CAT**) has only three letters, we have only enough data left in the buffer to do one final comparison. At this point, the function reads another chunk of data. In fact, the function could actually wait one more iteration before reading more data, by changing the line:

```
if (stop < length(mybuffer))
```

to use less-than-or-equal-to instead of less-than.

Let's see what remains of the buffer, the data that we have not yet checked.

```
K>> mybuffer(start: length(mybuffer))

ans =

AAC
```

We will look for that pattern shortly, to make sure it gets copied to the beginning of the buffer. Now we issue the **dbstep** command, to proceed by one more line.

```
K>> dbstep
89   mybuffer = [ mybuffer(start:length(mybuffer)), new_chars ];
```

The next line will move the data from the end of the buffer to the beginning, and append the next chunk of data (stored in the array **new_chars**). Next, let's see how much data was read. The length of the array `new_chars` will indicate this. While we are at it, we will examine the first 10 characters of the buffer, then "step" though the function by another line. This means that the line above that re-assigns `mybuffer` will take effect.

```
K>> length(new_chars)

ans =

    89

K>> mybuffer(1:10)

ans =

TTCCCAAGGC

K>> dbstep
90    total_chars_read = total_chars_read + chars_read;
```

Now we can re-examine the first 10 characters of the buffer. We should see the data left over from the first file read, followed by the data of the next file read.

```
K>> mybuffer(1:10)

ans =

AACCCCCTGT
```

We see here that the buffer has new contents, as we expect. Also, we see the previous data **AAC** start the buffer, as it should. Can we also verify that the rest of this data shown comes from the data read? We simply need to compare the buffer contents from above with the first seven characters of the `new_chars` variable.

```
K>> new_chars(1:7)

ans =

CCCCTGT
```

We visually inspect the data here to conclude that it is the same. We can continue on with the function, using the **dbcont** command.

```
K>> dbcont
87   [new_chars, chars_read] = fscanf(myfile, '%c', ...
K>> dbcont
>>
```

As the function call completed, MATLAB left debug mode and returned us to the familiar prompt.

But we had to issue the **dbcont** command twice. It stopped again, once we hit the line with the breakpoint. But why did the function execute that line again? The function reads the file for a new chunk of data every time it runs out of data. It gives up only when the number of characters read is zero. The second read brought in 89 characters, even though we asked for over 500. The function does not distinguish between successful reads, as long as it reads at least one character.

Are you still not convinced? One fault of the tests above is that the test data often repeats. How can we be sure that the **C** characters at the boundary between the first and second file read were not accidentally over-written? We will perform a quick check, by using a different data file, called "**DNA_example2.txt**." We make the new file by copying the old one, but with one exception. Using a separate file editor, we change the data starting at location 510 to be the alphabet, backwards. Let's test this out.

```
>> a = matgrep2('CAT', 'DNA_example2.txt');
pattern = "CAT"
filename  = "DNA_example2.txt"
K>> dbstep
89   mybuffer = [ mybuffer(start:length(mybuffer)), new_chars ];
```

With the original breakpoint in place, we step through to the next line. Now when we examine the last of the current buffer, we see the backwards alphabetic characters. We see the sequence continue in the beginning of the data from the file read. For comparison, we also show the old contents of the beginning of the buffer.

```
K>> disp(mybuffer(start: length(mybuffer)))
ZYX
K>> disp(new_chars(1:7))
WVUTSRQ
K>> disp(mybuffer(1:10))
TTCCCAAGGC
```

Now we take a step, which will execute the buffer re-assignment. Then we check the beginning of the buffer.

```
K>> dbstep
90    total_chars_read = total_chars_read + chars_read;
K>> disp(mybuffer(1:10))
ZYXWVUTSRQ
```

As we can readily verify, the old data and the new data are seamlessly put together in the new buffer.

3.10.6 Generating Random Data

We used random data to give us something to search. You may wonder how we created that data, and this section explains the process.

The goal is to have random DNA data, so we want to generate strings with only the letters A, C, G, and T. Technically, we will generate a pseudo-random sequence of values because the same sequence could be repeated if we use the same seed twice. MATLAB provides a **rand** function that returns a sequence of numbers between 0 and 1 with uniform distribution. We pass a parameter of **SIZE** to it, to specify how many random numbers we want. This variable is one that we define. Since we want an array instead of a matrix, we specify **rand(1,SIZE)**, as in the lines below.

```
SIZE = 600;
gene = floor(rand(1,SIZE)*4);
```

Now, we want a random sequence of characters, but our **rand** function returns a sequence of floating-point numbers. To turn the results into something we can use, we first multiply by 4 since we have four different letters. This way, all the numbers generated will be between 0 and 4. Then we use the **floor** function to make the numbers exactly 0, 1, 2, or 3. There should not be any values of 4, since the **rand** function returns values up to (but not including) 1.

Next, we map the integer values to letters. There are several ways to do this; we will use a series of **if** and **elseif** statements to determine the values, and use a **for** loop around it to repeat the process for each value.

```
for n=1:SIZE
    if (gene(n) == 0)
        str(n) = 'A';
    elseif (gene(n) == 1)
        str(n) = 'C';
    elseif (gene(n) == 2)
        str(n) = 'G';
```

```
        else
            str(n) = 'T';
        end
    end
```

From the code above, we can take each value of **gene(n)** $(0, 1, 2, \text{or } 3)$ and set **str(n)** accordingly to A, C, G, or T. It is simple and straight-forward, and this approach works for a variety of programming languages. While this will do the job, there are other ways. For example, we could use a **switch** statement instead, though this is more a matter of personal preference. But below we accomplish the same task, using the power of MATLAB.

```
letters = 'ACGT';
str = letters(gene+1);
```

Normally, we would expect to access **letters** with a single index. But MATLAB allows the index to be an array, and returns an output for each index. We add 1 to each index value, since MATLAB requires all indices to start at 1. In fact, this method should be much faster since MATLAB gives optimal performance working with ranges. A simple test for a **SIZE** of 100000 took 78.9 seconds with the **if..else** approach, but only 0.04 seconds with the above method!

Now that we have our random data, we need to save it to a file. First, we will open a file for writing. When given the parameter **'w'**, the **fopen** command will create the file if it does not exist. If it does exist, it will overwrite that file.

```
% Open the file to write
filename = 'DNA_example.txt';
myfile = fopen(filename, 'w');
```

We do not have to specify the file name separately, as shown above, but this way makes it easy to change. Variable **myfile** identifies the file for later commands. We could have several files open at the same time, so we use the file identifiers to keep track of them. File operations sometimes fail, so we should check to see that the call to **fopen** worked. MATLAB returns a -1 value for the file identifier in that case.

```
% Check for an error
if (myfile == -1)
    error(sprintf('Problem opening the file "%s".', filename));
    return
end
```

Next, we write the data to the file. Again, we check for error. The `fwrite` function returns the number of characters written. If the number written does not match the number we tried to write, then we know it encountered a problem. For example, if there is not enough disk space left to hold the file, then we will have such an error.

```
% Write the data
num_written = fwrite(myfile, str, 'char');
if (num_written < length(str))
    error(sprintf('Not all data was written to file "%s".', ...
        filename));
end
```

Finally, we need to close the file, telling the operating system that we are finished with the file for now. Again, we check the return value of the function, in case a problem exists.

```
% Now close the file
if (fclose(myfile) == -1)
    error(sprintf('Problem closing the file "%s".', filename));
end
```

With the file written, we end this program. Below appears the program, as a whole.

```
% make_data.m
%
% Create a random sequence of genomic data, and
% write it to a file.
%

% Say A=0, C=1, G=2, and T=3
SIZE = 600;
% Get an array of random numbers, between 0 and 3.
gene = floor(rand(1,SIZE)*4);

% Convert the fast way
letters = 'ACGT';
str = letters(gene+1);

% Now write the data to a file.
% Open the file to write
filename = 'DNA_example.txt';
myfile = fopen(filename, 'w');
% Check for an error
```

```
if (myfile == -1)
    error(sprintf('Problem opening the file "%s".', ...
        filename));
    return
end

% Write the data
num_written = fwrite(myfile, str, 'char');
% Check for an error
if (num_written < length(str))
    error(sprintf('Not all data was written to file "%s".', ...
        filename));
end

% Now close the file
if (fclose(myfile) == -1)
    error(sprintf('Problem closing the file "%s".', filename));
end
```

With the file search and data generating programs, we have seen how to include file input and output with MATLAB programs.

3.11 ANALYZING A CAR STEREO

This project comes from a practical problem. Listening to music in a particular car, we noticed that some songs sounded great while others sounded "fuzzy." Certain songs just do not sound right. But music listening is a subjective experience; how can we determine if a problem really exists? We have a hunch that one of the car's stereo speakers could have a bad connection, or possibly has damage. This would explain why certain songs sound fine while others do not.

To test this theory, we develop a frequency sweeping sound file, also called a chirp. Yes, like the program to follow, birds make a chirp sound, too. We start with a very low frequency, and gradually work our way up. Since it could be a problem with one channel, we will keep the channels separate. Also, we would like to know what frequencies cause the problem, if in fact we confirm our suspicion. Once we create the sound files, we will write them to a CD. The sound files should be long enough to notice a problem (e.g., silence), but we have no need to store the more than 70 minutes' worth of sound that a standard audio CD can hold.

Let's create a short audio sequence at an audible frequency. First, we set up some parameters, according to CD rates.

```
sampling_rate = 44100; % CD rate
bits_per_sample = 16;  % CD rate
```

We need the **bits_per_sample** when we store the sound data (i.e., with the **wavwrite** command).

Now we set up other variables for this particular signal. We use 2600 Hz as our example frequency, and plan for the sound to last for two seconds.

```
frequency = 2600; % Hz
number_of_seconds = 2;
```

Now we create the sound data. Variable **t** specifies the time, or duration of the sound.

```
t = 0:(1/sampling_rate):number_of_seconds;
data = 0.9*cos(2*pi*frequency*t);
```

Now let's hear the sound file.

```
sound(data, sampling_rate);
```

It may not be very pleasant, but we have sound! We can save it to a file with the **wavwrite** command, to create **2600Hz.wav**.

```
wavwrite(data, sampling_rate, bits_per_sample, '2600Hz.wav');
```

Notice that we did not specify how loud the audio should be. With **.wav** files, the sound samples should be greater than −1 and less than +1, which explains why we multiply the cosine function with a constant 0.9. Yes, we could use 0.99 or some other constant instead, as desired. We have some control over the loudness in that the larger values' magnitudes, the louder it will sound. For example, if we play a song where the minimum and maximum values are, say, −0.5 and +0.5, it would not sound as loud as the range from −0.9 and +0.9, the minimum and maximum that we use above. Of course, the listener has control over how loud something sounds, and can always turn the volume up or down.

If you play the sound and listen carefully, you should hear it equally well from either the left or right speaker. The **sound** function does that for us. But we can create a two-channel signal by giving our variable **data** two columns. We will split the data into two channels, playing the frequency in one channel only, then switch it to the other. The next few lines may look a bit complex, but we specify ranges. We create a variable called `first_half` that contains the indices for the first half of our `data` variable. A similar variable called `second_half` likewise contains indices corresponding to the other half of our data.

```
first_half = 1:round(length(data)/2);
second_half = round(length(data)/2):length(data);
```

Now we can assign the first half of the data to channel one.

```
data2(first_half, 1) = data(first_half);
```

Before we go further, let's check out some things about the new variable, `data2`.

```
>> size(data2)

ans =

       44101              1
```

This new variable contains even less information than the original `data` variable. Let's examine its contents, or at least the first few values.

```
>> data2(1:10,:)

ans =

       0.9000
       0.8390
       0.6641
       0.3991
       0.0800
      -0.2499
      -0.5460
      -0.7680
      -0.8858
      -0.8834
```

These values are typical; they are in the range of $(-1,+1)$, like we expect. The first value would have been 1, had we not scaled it to 0.9.

Next we put the second half of data in channel two.

```
data2(second_half, 2) = data(second_half);
```

The second line does a bit more than the first. Since we specify data for a second dimension, MATLAB fills in the blanks with zeroes. To see this, we re-examine the size.

```
>> size(data2)

ans =

      88201             2
```

Matrix **data2** does not contain twice as much data; it contains four times as much! We doubled the length, but we also doubled the width. Now let's see the first few values again.

```
>> data2(1:10,:)

ans =

     0.9000            0
     0.8390            0
     0.6641            0
     0.3991            0
     0.0800            0
    -0.2499            0
    -0.5460            0
    -0.7680            0
    -0.8858            0
    -0.8834            0
```

The values in column 1 match what we saw above. Column 2 contains zeros for the corresponding locations. That is, a speaker playing channel 2 will be silent for these values. Later on, channel 1's speaker will be silent. We verify this by examining some values of **data2** well into the second half.

```
>> data2(60000:60010, 1:2)

ans =

        0     -0.5561
        0     -0.7746
        0     -0.8879
        0     -0.8809
        0     -0.7543
        0     -0.5254
        0     -0.2252
```

```
0      0.1055
0      0.4220
0      0.6811
0      0.8479
```

As we see, channel 1 contains zeros while channel 2 has data from the frequency of interest.

Now we can test the two channels. Play the sound in variable **data2** like we did above.

```
sound(data2, sampling_rate);
```

Listening carefully, you will hear the tone on one speaker, then the other.

We next create a sweep of frequencies, where we gradually increase frequencies over time. Since CDs use a sampling rate of 44,100 samples per second, the maximum frequency is 22,050 Hz. At the other end of the scale, humans typically do not hear frequencies lower than 20 Hz. Thus, we want to have a sweep of frequencies from 20 Hz to around 20 kHz.

A typical stereo speaker looks like the drawing in Fig. 3.8. We see that several speakers comprise what we would call one of (the two) speakers. The reason why we have several speaker cones instead of just one has to do with their responses; the largest one (woofer) does a good job of faithfully reproducing low frequency sounds, while the smallest one (tweeter) reproduces high frequency sounds well. For the car stereo, we suspect that one of these cones does not perform well, perhaps due to a loose connection or damage. By sweeping through the audible frequencies, we should be able to notice a silence, and narrow down exactly which speaker has the problem.

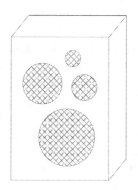

FIGURE 3.8   **A typical stereo speaker.**

But humans do not differentiate between high frequencies very well. We can take advantage of this by grouping the higher frequencies together, by changing between them quickly.

```
sampling_rate = 44100;
start_freq = 2000;
stop_freq = 3000;
number_of_seconds = 2;
t = 0:(1/sampling_rate):number_of_seconds;
```

We make our increment so that frequencies will be the same length as **t**.

```
increment = ((stop_freq - start_freq)/length(t));
frequencies = start_freq:increment:stop_freq-increment;
sweep = 0.9*cos(2*pi*frequencies.*t);
sound(sweep, sampling_rate);
```

Listening to the **sweep** signal, we hear that the frequency increases over time. It sounds like a slide-whistle. If you have a good ear for frequencies, you may even notice that it sounds higher pitched than it should, at the end. The code above has a bug; it actually generates frequencies between 2000 Hz and 4000 Hz, and we will discuss the reason why shortly.

But for now, we focus our attention on making a function to provide sweep data. The code above consistently gives us a sweep of twice the specified bandwidth, so we can fix it by setting the stop frequency to be half the requested bandwidth. For example, with a start frequency of 2000 Hz and a stop frequency of 3000 Hz, the bandwidth is $3000 - 2000 = 1000$ Hz. So we create a "real" stop frequency of $2000 + 1000/2$ Hz.

We need to do a few things to finalize code that generates frequency sweeps. First, we should "abstract out" the code above. In other words, we should turn that code into a function, where we pass a few parameters, and have it return the sweep signal.

Which variables do we need for this function? We will want to generalize the function, so that we can use any start and stop frequencies. Also, we need to know the number of seconds. That way we can indirectly control how quickly we go from one frequency to another. We do not need to pass variable **t**, since we can calculate it within the function. But we do need to know the sampling rate. Variables **increment** and **frequencies** can be calculated from the parameters we have already mentioned, as can the output data **sweep**. It is a good idea to comment out the **sound** command. Once we are satisfied that the code works, we will use it over and over again. Playing each sweep will cause the function to slow down considerably, and the user can always call the **sound** function himself when desired. We can be helpful by leaving it in the program as one of the header comments, so the user will see it as part of the **help**.

We make a subtle change on the size of variables **t** and **frequencies**, effectively shortening them by one element. This has no dramatic effect, but it does make the arrays exactly in intervals of seconds. For example, if we have 44100 samples per second, it makes sense to have 44100 samples after one second. Since we start at 0 seconds, we have 44101 samples when we stop at 1 second, so we will stop it one sample earlier.

The **freq_sweep** function appears below.

```
% freq_sweep.m
%
% Generate a frequency sweep, a signal where the frequency
% gradually gets larger.
%
% Usage:
%    sweep =  freq_sweep(start_freq, stop_freq, ...
%        number_of_seconds, sampling_rate)
%
% Example:
%    [sweep] = freq_sweep(2000, 3000, 10, 44100);
%    sound(sweep, 44100);
%

function [sweep] = freq_sweep(start_freq, stop_freq, ...
    number_of_seconds, sampling_rate)

% Check our assumption
if (start_freq == stop_freq)
    error('start frequency must not equal stop frequency.');
    return
end

% Find the range of frequencies
bandwidth = stop_freq - start_freq;
% Make the real stop frequency at half the bandwidth.
% (This is a complex issue - but it works this way!)
real_stop = start_freq + round(bandwidth/2);

% simulated time
t = 0:(1/sampling_rate):number_of_seconds-(1/sampling_rate);
```

```
% Make increment so that frequencies will be
% the same length as t
increment = ...
    (real_stop - start_freq)/(sampling_rate*number_of_seconds);

% Make the frequency get larger over time
frequencies = start_freq:increment:real_stop-increment;

% Since variables "frequencies" and "t" have exactly the same
% size, we can multiply them together point-by-point. This
% gives us the sweep data.
sweep = 0.9*cos(2*pi*frequencies.*t);
```

We can test it out, like the header comments suggest.

```
sampling_rate = 44100;
[sweep] = freq_sweep(2000, 3000, 10, sampling_rate);
sound(sweep, sampling_rate);
```

As expected, we hear a frequency sweep from 2000 Hz to 3000 Hz. But how can we be sure that these are the frequencies generated? Examining the spectrum will answer that question. The following lines show the frequency plot that we see in Fig. 3.9, with frequencies (x-axis) versus amplitude (y-axis).

```
SW = fft(sweep);
half = 1:round(length(SW)/2);
plot(half*sampling_rate/length(SW), abs(SW(half)), 'k')
axis tight
```

Briefly, the code above converts the time-domain data to the frequency domain. The theory behind this code is beyond the scope of this text. For more information about how these lines work, consult a book on digital signal processing.[2]

The graph in Fig. 3.9 shows that we have mostly zero amplitude, except for a narrow band. Since the x-axis has a scale of 10^4, point 0.4 along the bottom corresponds to 4000 Hz. Careful examination reveals that we have frequencies between 2000 and 3000 Hz.

3.11.1 A Fun Sound Effect

Interestingly, we do not have have to give a start frequency smaller than the stop frequency. Consider what happens when we do the following.

---

[2]M. Weeks, *Digital Signal Processing Using MATLAB and Wavelets*, Infinity Science Press, 2006.

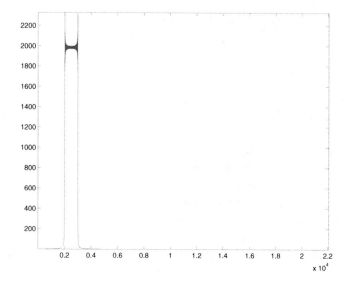

FIGURE 3.9 **Spectrum from the sweep function.**

```
[sweep2] = freq_sweep(3000, 2000, 10, 44100);
sound(sweep2, 44100);
```

Playing **sweep2** reveals that it sounds like a backwards version of **sweep**, since the frequency falls over time. It is much like the sound effect used for a falling object, such as in a war movie when an airplane drops a bomb.

3.11.2 Another Fun Sound Effect

Here is another fun sound effect, assuming that we still have variables **frequencies** and **t** defined.

```
sweep2 = 0.9*cos(2*pi*round(frequencies).*t);
sound(sweep2, sampling_rate);
```

We produce a sweep function, but one with whole numbers all the frequencies. This step between frequencies means that the frequency function is not linear. It results in an interesting sound, with a more complicated spectrum. We can see the spectrum with the code below.

```
SW2 = fft(sweep2);
half = 1:round(length(SW2)/2);
plot(half,abs(SW2(half)),'b')
```

This spectrum shows frequency content spread out over all frequencies, instead of neatly confined to a small area.

3.11.3   Why Divide By 2?

Examining the code in **freq_sweep**, you may have noticed that it calculates a bandwidth and a variable called **real_stop** that we use in place of the given stop frequency. Variable **real_stop** has a value half way between the given starting and stopping frequencies. This has to do with the argument to the **cos** function, normally a linear function of time. When we create a frequency sweep, though, we change the frequency over time. In that sense, the argument becomes a function of time squared (i.e., the frequency (f) itself can be expressed at a function of t and we already multiply the frequency by t in the formula $cos(2\pi ft)$).

We can model the frequencies in a different way. First, remember our example values.

```
sampling_rate = 44100;
start_freq = 2000;
stop_freq = 3000;
number_of_seconds = 2;
```

Now we re-examine the code from earlier, with the bug.

```
t = 0:(1/sampling_rate):number_of_seconds-(1/sampling_rate);
% Make increment such that frequencies
% will be the same length as t
increment = ((stop_freq - start_freq)/length(t));
frequencies = start_freq:increment:stop_freq-increment;
sweep = 0.9*cos(2*pi*frequencies.*t);
```

One way to look at this is to examine the original sweep function.

```
sweep = 0.9*cos(2*pi*frequencies.*t);
```

Our model of the **frequencies** shows that the above line can be written in the following, equivalent, fashion.

```
bandwidth = (stop_freq - start_freq);
sweep2 = 0.9*cos(2*pi*start_freq*t + ...
    2*pi*bandwidth/(number_of_seconds)*t.^2);
```

To show that the two signals **sweep** and **sweep2** are equivalent, we can calculate the maximum difference between the two.

```
>> disp(max(abs(sweep - sweep2)))
    1.2113e-11
```

We see that there is no difference between any two points of these arrays in the first 10 digits of precision.

The `length(t)` represents the total length of the signal, the duration of the signal multiplied by the number of points per second. We can think of a new variable `n` as ranging equivalently from `0:number_of_seconds * sampling_rate` -1, with increments of 1. So range `n` is the same as `t`, except that values in `n` are whole numbers. In other words, $t = \frac{n}{\texttt{sampling\_rate}}$.

```
n = 0:number_of_seconds*sampling_rate -1;
```

Our frequencies range from the given start frequency to the stop frequency. We create the variable `frequencies` to gradually go from one to the other, based on value of `increment`. The number of elements in `frequencies` should match the number of elements in `t` (or `n`). Our `increment` variable adds a tiny bit to the start frequency to get the second element of `frequencies`. Then it adds a tiny bit more to get the third value of `frequencies`. Each value in the array takes the previous value and adds the small `increment` to it. In this way, each value depends not only on the previous value, but on *every* value before it. We could use the above range `n` to find an equivalent set of frequencies.

```
g(n+1) = start_freq + n*increment;
```

To verify that the arrays `g` and `frequencies` are the same, we can see the maximum difference between them.

```
>> max(abs(g-frequencies))

ans =

    4.5475e-13
```

A difference that tiny can be explained by round-off error. In order to find an arbitrary frequency, say, `f(123)`, we need to know the previous value, in this case `f(122)`. If we do not have that, we can start with the very first frequency (`start_freq`) and add the product of the increment and the index. In other words, the above MATLAB command defining `g` works for the parameter `n` whether it is a scalar or an array. Now we switch the index from the integer `n` to the real value `t`. The frequency varies over time; as we get to the "end," or maximum value for `t`, the frequency function should give us the stop frequency. Recall our relation between `t` and `n`:

$$t = \frac{n}{\texttt{sampling\_rate}}.$$

Replacing **n** with **t**, we define our frequency function **f** in a similar way as above. We can make **t** arbitrarily small. From this point forward, we will use

$$\text{increment} = \frac{\texttt{stop\_freq} - \texttt{start\_freq}}{\texttt{sampling\_rate}}$$

and

$$f(t) = \texttt{start\_freq} + \texttt{increment}\, t.$$

When we consider the argument to the cosine function as a function of time, things get even more interesting. We can call it $\theta(t)$.

$$\theta(t) = \int 2\pi f(t)\, dt$$

We can find $\theta(t)$ for any given t.

$$\theta(t) = \int 2\pi\, \texttt{start\_freq}\, dt + \int 2\pi\, \texttt{increment}\, t\, dt$$

We know the initial conditions: that the increment is zero at $f(0)$ and that $\theta(0) = \texttt{start\_freq}$. Solving the integral gives us:

$$\theta(t) = 2\pi\, \texttt{start\_freq}\, t + 2\pi\, \texttt{increment}\, \frac{t^2}{2}.$$

The point is that the increment should only be half as much as what we might expect, due to the $t^2/2$ term that follows it. Now, we can substitute in $\theta(t)$ above into MATLAB code. First, we recall the example values to use.

```
sampling_rate = 44100;
start_freq = 2000;
stop_freq = 3000;
number_of_seconds = 2;
bandwidth = stop_freq - start_freq;
```

Now we calculate the time variable **t** and the **sweep**, using $\theta(t)$ from above.

```
t = 0:(1/sampling_rate):number_of_seconds-(1/sampling_rate);
sweep = 0.9*cos(2*pi*start_freq*t + ...
    2*pi*bandwidth/(2*number_of_seconds)*t.^2);
```

When we use this version of the code, we find the frequency plot to be the same as what we saw in Fig. 3.9. In other words, we might prefer the code that creates the range **frequencies**, due to clarity or efficiency, but it has the same effect as the code above. Either way, we see the need for the division by 2 from the above analysis.

3.11.4 Stereo Test Conclusion

So what was the outcome of the test? To create a CD of sweep sounds, we need a program to call `freq_sweep`, then we generate an appropriate file name, and then we write the data to the file. The program `create_sweep_files.m` (included on the CD-ROM) does this.

Instead of manually creating each data set, we set up a variable called `myranges` that defines the starting frequencies.

```
myranges = [   20,  100,  200,   400,   600,   800,  1000, ...
             1400, 1800, 2200,  2800,  3400,  4000,  5000, ...
             6000, 7000, 9000, 11000, 13000, 16500, 20000];
```

Given each starting frequency, we create a sweep ending at the next starting frequency. For example, the first sweep will be from 20 Hz to 100 Hz. Notice how the difference between frequencies gets larger as the frequencies increase. Humans cannot distinguish high frequencies from each other very well. We take advantage of this fact to limit the length of the test signals for high frequencies.

We use the `freq_sweep` function to find a frequency sweep between two frequencies. The array `myranges` holds these frequencies, we simply need to indicate which one (and its neighbor). Suppose that we choose variable n, not to be confused with the n from the last section, to indicate the current starting frequency. Since we will repeat these steps for each starting frequency, n can be defined as the index to a loop. Then we make the call, as shown below.

```
sweep = 0.9*freq_sweep(myranges(n), myranges(n+1), ...
    number_of_seconds, sampling_rate);
```

We have to be careful not to let n be as large as the number of values in `myranges`, otherwise the code above will try to access a value that does not exist.

Next, we can plot the spectrum. While this is not necessary, it gives us a visual check of the frequency content of the data we create.

```
SW = fft(sweep);
half = 1:round(length(SW)/2);
plot(half*sampling_rate/length(SW), abs(SW(half)), 'b')
info = sprintf('Sweep from %d to %d.', ...
    myranges(n), myranges(n+1));
title(info);
```

The code above shows the spectrum, along with a descriptive title.

Now we create the file name. Assume that the variables n and channel_number are already defined. As noted, we use n as an index to run through all starting frequency values in myranges. But we want to do that twice, once for each channel. So channel_number will be another loop index. To use it in a file's name, the sprintf command combines it with text.

```
channel = sprintf('ch%d', channel_number);
filename = sprintf('%s_sweep%d.wav', channel, n);
disp(sprintf('Writing file "%s".', filename));
```

The last line shows us the created file name. You may want to try those three lines out yourself, with example values for n and channel_number. For instance, MATLAB displays the following line when n equals 4 and channel_number equals 2.

```
Writing file "ch2_sweep4.wav".
```

When we have data for channel 1, we still need to provide data for the second channel. Since we want channel 2 to be silent, we use the zeros command to set all values to 0. We can use the following function call to make a column of zeros for that channel.

```
zeros(length(sweep), 1)
```

For stereo sound, we need a column for each channel. The code below creates a matrix with two columns; the sweep data in the first column, and zeros in the second column.

```
[sweep.', zeros(length(sweep), 1)]
```

Now we can write the stereo data to a file with the wavwrite command.

```
wavwrite([sweep.', zeros(length(sweep), 1)], ...
    sampling_rate, 16, filename);
```

This will write a .wav file with our data. Before executing that line, we should check the channel number. If we are working on channel 2, we need a similar line with the sweep and zeros switched. So we put the code above in an if statement, with a similar line in the else block to take care of the second channel.

The program create_sweep_files.m incorporates the above tasks. You can use it to create the same data files as used below.

Once the files are created, we can import them into a program supplied with the computer's operating system, and burn an audio CD. For example, Apple's iTunes can create audio CDs, but most operating systems will ask the user if he wants to burn a CD when it detects a blank one in the CD drive. After creating the CD, the car's stereo was tested.

The car has six speakers total; three on the driver's side and three on the passenger's side. The largest ones, the woofers, are built into the respective doors. A medium and a small speaker are connected to each other, each set on either side of the dashboard. This makes it difficult to isolate the problem; that is, listening to the one side of the dashboard we cannot easily differentiate the sounds from the smallest speaker and those from the medium speaker. Speakers can be covered with fabric (e.g., a towel), but this does not completely block the sound from that speaker.

To help analyze the sound, we use a sound level meter. However, this device has sensitivity to position. The closer we hold it to a speaker, the louder the sound registers. Readings could be off by several decibels as a result, but this is not a concern. For the purpose of this test, we can test each speaker in a binary fashion. Either the speaker works, or it does not.

The first channel corresponds to the driver's side. Volume was set to a value of 6, high enough to be heard, but low enough to not scare the driver when turning on the car. The sound level meter was held about 2 inches from the speaker under observation. The windows were rolled up, the engine was off, and the car was parked in a relatively quiet environment. The sound level meter recorded values typically between 60 dB and 90 dB. Sounds below 60 dB were below the sensitivity of the meter. To put this in perspective, a ceiling fan registers over 60 dB when the meter is within 5 feet.

The test produced two somewhat disturbing observations. First, track 20, which contains frequencies above 16,500 Hz, was inaudible. However, the sound level meter also did not register the sound, so this says more about the speakers than the listener. Second, the relatively low volume of the stereo still put out over 90 dB of sound, enough to cause hearing damage over time. Of course, the relative position of the listener could account for this. In other words, a person is not likely to wedge his head between the dashboard and windshield to listen to the stereo.

The plastic mesh that protects the cone of each speaker vibrates when the speaker produces sound. This provides a helpful clue, to support our results.

After measuring the sound by holding the sound level meter above the speakers, it was determined that the driver's side speakers worked well. Turning to the passenger's side, we found that the sound tracks 21 through 25 were audible, but mostly from the woofer. Tracks above 28 were also audible, from the tweeter. Tracks 26 and 27, where the mid-range speaker should have handled (around 800 Hz to 1400 Hz) were audible but not loud. Further measurements revealed that the woofer and tweeter were responsible for the sounds heard. In other words, the passenger's side mid-range speaker was not functioning. This was further confirmed by setting the stereo's balance to be all on channel 2. Tracks 1 through 20 were not audible under this setting, just as we would expect.

Examining the speaker did not reveal any obvious damage, and the connections looked to be good. However, the speaker connects to the stereo in an inaccessible area of the dashboard, and one of these connections could be the problem. The mid-range speaker did not vibrate under the sweep signals, and the sound measured above the speaker was the same as sound measured above the dashboard on the other side of the tweeter. In other words, the tweeter provided sound while the mid-range speaker did not. Therefore, the conclusion is that the passenger's mid-range speaker does not work, and should be replaced or at least reconnected.

3.12   DRAWING A LINE

Eventually, we will draw a three-dimensional cube. To do that, we need to be able to solve a simpler problem: drawing a square of one of the cube's sides, which may have a diamond shape from our viewpoint. In order to draw such a shape, we need to solve an even simpler problem first, drawing a line. We discover how to draw a line on an image in this section.

3.12.1   Finding Points Along a Line

Given two points by the coordinates $(x1, y1)$ and $(x2, y2)$, we will draw a line segment that connects the two. On paper, this is easy: we draw the points, put a straight-edge ruler between them, and draw the line segment. But what if we had to do this without making a graph, if we had to give a list of (integer) coordinates of points between the two points, based on their coordinates?

Suppose we have a trivial example, where we need the points on the line segment between points (1,2) and (4,5). The points (2,3) and (3,4) lie on that segment, which we can verify easily. But to find these points systematically, we need a few observations. First, we recall the equation $y = mx + b$ that specifies a line. We must have at least two points to make our line, so we have two equations:

$$2 = m \times 1 + b$$

and

$$5 = m \times 4 + b.$$

Since we have two equations with two variables, we know we can solve for these variables and find the slope m and the y-axis intercept b. From the first equation, we get $b = 2 - m$. Then, we substitute $2 - m$ for the b value of the second equation. This gives us

$$5 = m \times 4 + 2 - m.$$

We can solve this for m by re-arranging the terms.

$$5 - 2 = m \times 4 - m$$
$$3 = m \times (4 - 1)$$
$$3 = m \times (3)$$
$$m = \frac{3}{3} = 1$$

Now we can use either of the line equations and substitute our slope, m with the value one. Let's use the first equation.

$$2 = 1 \times 1 + b$$
$$2 = 1 + b$$
$$2 - 1 = b$$
$$b = 1$$

Now we can use our slope and y-axis intercept with the line equation,

$$y = 1 \times x + 1.$$

We can also represent this as a function of y, with the equation

$$x = \frac{y - 1}{1}.$$

Naturally, we do not need to divide by 1, but the term in the above equation reminds us that the equation as a function of y has a division by the slope. So if we have either the x or the y coordinate from a point on this line, we can find the other. To get points on the line segment between the two given points, we already know the range of x or y coordinates to use. We simply form an array using the given y coordinates as the start and end points, and calculate the corresponding x values according to the line equation.

In MATLAB, we can create an array with `y1:y2`. This has a problem since it assumes that `y1` is less than `y2`. Instead, we will compare the two values and make a step size of either 1 or -1 between them. For example, `2:1:5` generates the array `2, 3, 4, 5`. As another example, consider `5:-1:2` that generates the array `5, 4, 3, 2`. The order does not matter, as long as we have a correct step size.

The code below will find the x values when given the y values.

```
>> y = 2:5

y =

    2     3     4     5

>> x = (y-1)/1

x =

    1     2     3     4
```

Here, the x and y values happen to be integers. For our purposes, we want to make sure that they are integers since we will use them to index an image. We will do this using the `round` function.

Another thing to consider is whether we should find x values based on the y values, or the other way around. Actually, it depends on the line. For a diagonal line with equal rise and run, it does not matter which one we use. In that case, we will generate the same number of points with either set of values. But consider a line that passes through the points (0,0) and (100,2). If we use the y values to define a range, it would be 0, 1, 2. Plugging these into the line equation gives us the corresponding x values, but only adds one point. In other words, we would specify this line segment with three pixels instead of 101. A better solution for this example would be to use the x values to generate y values. How do we know when to use the x values to generate points, and when to use the y values? We want to generate the most points that we can, so we check for the values with the largest range.

3.12.2 Coding the Solution to Points Along a Line

Now that we have a good problem definition, we can solve it with MATLAB code. We want to find the points on a line segment between two given points, taking into account the issues presented above. We start with a function definition.

```
function [newx, newy] = ...
    points_on_line(point1x, point1y, point2x, point2y)
```

The inputs are two sets of coordinates, specifying two points. Based on these points, it finds the slope and the y-intercept. A first attempt appears below.

```
slope = (point2y - point1y)/(point2x - point1x);
b = point1y - slope*point1x;
```

But before defining the slope, we should check the difference between the two x coordinates. A difference of zero indicates a vertical line, and will produce an infinite slope. But the calculation involves a divide by zero. MATLAB handles these fairly well, advising us of the divide by zero, and assigning `Inf` to the value of **slope**. However, we can avoid the warning that division using a zero denominator generates by simply giving the slope a very large value. Assuming that the coordinates will be relatively small, this will work fine. The **if** statement below checks the denominator, and makes the slope a very large value instead of infinite, when needed.

```
if (point2x - point1x ~= 0)
    slope = (point2y - point1y)/(point2x - point1x);
else
    slope = 100000000;
end
b = point1y - slope*point1x;
```

Next, it chooses to work with the x values of y values depending on which one contains more points. The partial **if** statement below determines the increment when the y coordinates have a greater difference than the x coordinates. We will have similar code in an **else** block to deal with the case when x coordinates define a larger range.

```
if (abs(point2y - point1y) > abs(point2x - point1x))
    if (point1y < point2y)
        inc = 1;
    else
        inc = -1;
    end
```

The increment will either be a positive one or a negative one. Technically, a negative one would be called a decrement. Once we have this, we can determine the range of *y* coordinates.

```
newy = round(point1y):inc:round(point2y);
```

Notice that the indentation reminds us that this line of code belongs to the first **if** statement. With the range based on the *y* coordinates, we can generate the *x* coordinates that go with them.

```
newx = round((newy - b)/slope);
```

You may recognize that the command above uses a modified version of the line formula to determine the *x* coordinates. After this, the function returns the new points. The **else** block follows, with similar code for generating *y* coordinates from a range of *x* values.

 The function **points_on_line.m**, on the CD-ROM, finds the points on a line segment between two given points. The inputs are two sets of point coordinates, four values in total. Based on these points, it checks for divide by zero, and finds the slope and the y-intercept. Then it decides on the increment (or decrement), creates the list of *x* (or *y*) coordinates, and finds the other coordinates. Finally, it returns the points as two sets: an array of *x* coordinates and an array of *y* coordinates.

Below, we test this function. First, we try something easy that we can readily verify.

```
>> [x_list, y_list] = points_on_line(3, 4, 8, 9)

x_list =

     3     4     5     6     7     8


y_list =

     4     5     6     7     8     9
```

The *x* values range from 3 to 8, while the *y* values range from 4 to 9. Since this line segment is a diagonal with a rise and run of one each, all points happen to be integers. You can verify this with graph paper; drawing a line segment between these points means that the line passes through the intersection of the grid lines on the graph. We see from the example above that we get all points between (3, 4) and (8, 9). Now let's try something a bit more complex.

```
>> [x_list, y_list] = points_on_line(0, 0, 2, 4)
```

```
x_list =

    0    1    1    2    2


y_list =

    0    1    2    3    4
```

We see in this example that we get five points. But suppose that we just got lucky this time; we try it once more, switching the *x* and *y* coordinates for the second point.

```
>> [x_list, y_list] = points_on_line(0, 0, 4, 2)

x_list =

    0    1    2    3    4


y_list =

    0    1    1    2    2
```

Here, we have the same set of points as before, only with the *x* and *y* coordinates reversed. This demonstrates that we get the maximum number of points along the line, as desired.

3.12.3 Drawing the Line

Now that we have points along a line, how do we draw it? Below we have a simple method. First, we need a background image. You can think of an image as nothing more that a matrix, where all elements have values from 0 (black) to 255 (white). This corresponds to grayscale, often used because you can demonstrate the idea without attending to the extra details for a color image. We will make a 512 × 512 pixel white image to start. We create a matrix of ones, with each value multiplied by 255.

```
MaxRow = 512;
MaxCol = 512;
testimage = 255*ones(MaxRow, MaxCol);
imshow(testimage);
```

We include the `imshow` command in the above code. The image is not yet interesting, but it gives us something to look at. Second, we choose some example end points, and find the coordinates in between them using **points_on_line**, defined above.

```
[x_list, y_list] = points_on_line(68, 66, 431, 440);
```

Third, we replace the values in our matrix with values that will stand out. We reshow the image after we place the points.

```
for k = 1:length(x_list)
    testimage(y_list(k), x_list(k)) = 0;
end
imshow(testimage);
```

Running the code above, you should see a black line on a white background.

Now let's try a more complex example. After drawing the line above, we can add a vertical one, a horizontal one, and a diagonal one at about a right angle. These lines are interesting because they include extremes of an infinite slope, and a zero slope. First, we draw a vertical line.

```
[x_list, y_list] = points_on_line(250, 67, 250, 410);
for k = 1:length(x_list)
    testimage(y_list(k), x_list(k)) = 0;
end
imshow(testimage);
```

Now we make a horizontal line.

```
[x_list, y_list] = points_on_line(90, 250, 400, 250);
for k = 1:length(x_list)
    testimage(y_list(k), x_list(k)) = 0;
end
imshow(testimage);
```

Finally, we draw another diagonal line.

```
[x_list, y_list] = points_on_line(110, 410, 330, 125);
for k = 1:length(x_list)
    testimage(y_list(k), x_list(k)) = 0;
end
imshow(testimage);
```

 Fig. 3.10 shows the results of our tests. We see all four lines. They do not meet in the center, since we did not ask for them to do that. As an exercise, what values could you give to the **points_on_line** function to have all four lines intersect at one point? If you want to experiment, the program **drawinglines.m** on the CD-ROM contains the code from above.

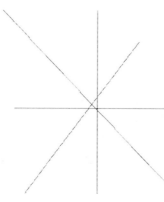

FIGURE 3.10   **Drawing lines on an image.**

3.13   DRAWING A FRAME

Now that we can draw a line on an image, let's take a step closer to drawing a three-dimensional cube by drawing an outline of it. To make this more challenging, we let the sides have colored tiles arranged in a 3×3 grid.

We need to know how to dimension it. We cannot make all of the squares the same size, since this does not take perspective into account. When drawing a scene, such as making a painting for an art class, objects are scaled in size depending on their distances. That is, imagine that you are painting what you see as you look down a street. If a car is right in front, you would paint it rather large. If a car by the same make and model is parked one block down the street, you might paint it only one-quarter the size. This is what we mean by perspective. Given the perspective of the user, some blocks of the cube are larger because they are closer. How can we get a good idea what these dimensions should be?

We can start by taking a digital picture of a cube, such as the cube made of paper (as in file `cube1.jpg`, on the CD-ROM). We could instead use a picture of something similar, like an office building, or a Rubik's Cube®(a toy made by Seven Towns Limited).[1]ither way, a small cube with 3×3 tiled sides preserves this idea without adding too many details. We will use the paper cube from the CD-ROM. Next, we can measure points on the picture with the `ginput` command. The user will be able to move the pointer over the

---

[1]Trademark by Seven Towns Limited.

image, and select a point by left-clicking the mouse. The commands below show the image, then allows the user to select a point on it. We can repeat the second line as many times as necessary, since it returns a pair of coordinates for a point each time.

```
imshow('cube1.jpg')
figure(1); [x1, y1] = ginput(1)
```

Another possibility is to use `ginput(16)`, which returns 16 points that we select. However we do this, we can select points that correspond to intersections of lines on the cube. Based on these points, we will draw an outline of the cube.

These points give us an approximation, but we can clean them up by adjusting them based on other points. For example, the two points between cube corners can be computed based on the line formed by those two corner points.

We use these points to define the dimensions of the cube. For the moment, we will draw the frame of the cube, the lines that define the different squares. First, we will start off with a default background.

```
flip = 0;
MaxRow = 700;
MaxCol = 700;
```

The dimensions of the cube (700×700) may seem arbitrary, but they come from the `cube1.jpg` picture. Originally, this picture was much larger, but it was cropped to get rid of as much of the background as possible. The variable `flip` controls how the cube will appear. We will want to show all sides of the cube, but can only see three at a time. The `flip` value lets us choose which sides we see. For the moment, only values 0 and 1 matter. Either the frame will appear as in the original image, or rotated so that the bottom face shows.

Next, we create the default background. We could use the `zeros` function to create a black background, but for now we will use `255*ones` that will create values of 255 for each pixel. This makes the whole image white.

```
im2 = 255*ones(MaxRow, MaxCol, 3);
```

Above we see the command to create a new image, called `im2`. Notice that we use a value of 3 for the third dimension. This image will appear in color, instead of grayscale. So the third dimension holds values for red, green, and blue components. We explain this more thoroughly later.

As mentioned above, we could pick the points based on the `ginput` command from the original picture. Below we have the point information, adjusted to look good, for the top face. We keep the x and y coordinates in separate arrays.

```
y1 = [  15,  38,  38,  73, ...
        69,  73, 108, 107, ...
       107, 108, 149, 155, ...
       149, 205, 205, 271];

x1 = [ 339, 262, 416, 147, ...
       339, 531,  33, 231, ...
       447, 645, 110, 339, ...
       568, 215, 463, 339 ];
```

These 16 sets of coordinates correspond to the 16 points on the top face (Face 1). These are ordered from top to bottom, and left to right. We will also create a set of 16 coordinates for the left face (**x2** and **y2**) and another 16 points for the right face (**x3** and **y3**).

When the flip is even, we use the original points. When variable **flip** has an odd value, we alter the points as needed. Essentially, we rotate the faces by 180 degrees when we have an odd flip value.

With these coordinates, we can draw an outline of the top face. To draw a line, we find the points along the line given two points. We use the **points_on_line** function, as described in a previous section.

```
[newx1, newy1] = points_on_line(x1(1), y1(1), x1(7), y1(7));
```

This command finds the points on the line created by points $(1, 1)$ and $(7, 7)$ that corresponds to the top-left edge. Now we can draw the line. Since we have a white background, we will draw black lines.

```
for k=1:length(newy1)
    im2(newy1(k), newx1(k), :) = 0;
    im2(newy1(k)+1, newx1(k)+1, :) = 0;
end
```

The **for** loop allows us to put all points from the line segment on the image. We assign values of the image array, **im2**, to be zero for all three colors. This will appear as black, and we do it point by point. You may notice that the assignment happens twice, the second time using coordinates that are incremented by 1. This second assignment makes the line look well defined, much like how you might move a pen back and forth to mark a line on paper. With a line definition of only a single pixel width, the line may not appear or may only partially appear, due to the software that controls how the image looks on screen. Doubling the width helps the line stand out.

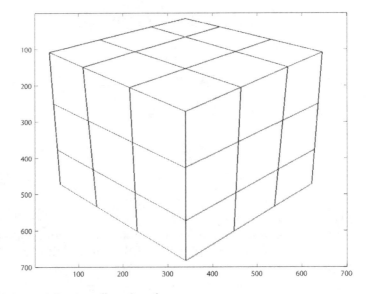

FIGURE 3.11   A frame outline of a cube.

 To finish drawing the outline of the top, we simply apply the code above for each line segment. This means four line segments that go diagonally from bottom left to top right, and four more diagonally from the top left to the bottom right. After we have the top completed, we next turn our attention to the left side, then the right side. The idea remains the same; we simply apply the **points_on_line** data to the image data. See the **draw_frame.m** program, supplied on the CD-ROM. It generates the image seen in Fig. 3.11. As expected, the figure shows a frame outline of the cube. Shortly, we will see how we can fill in the frame.

Here, we drew a line, then repeated it over and over again using the points that we obtained from a digital picture. It may not be clear where the vertex points come from, but they appear on a later figure. From Fig. 3.11, we observe a few things. First and foremost, the image looks like a cube! We were able to capture the perspective of a three-dimensional cube. On close inspection, you can see that the corner where the three sides meet appears larger than the other corners, as if it were closer to us. Also, the lines that define the left-most and right-most edges appear to go straight down, though they actually are not parallel with the axes that MATLAB includes on the figure. It looks symmetrical, which it should; the vertex points on the right were calculated from those on the left half of the image.

Next, we turn our focus to filling in the diamond shapes shown in the frame.

3.14   FILLING A DIAMOND SHAPE

To draw a cube as an image, we need to fill in the diamond shaped squares. Notice that we can fill a square or rectangle easily, with a command like the following.

```
myimage(100:120, 50:90) = 255;
```

The problem comes about when we try to generalize this. A square viewed from an angle does not appear as a square, but as a diamond. Consider a simple diamond shape: 1 pixel across the top, 3 pixels on the next row, then 5 pixels, then 3, then 1 again. To faithfully create this shape, we need to change some rows of pixels between a range of columns, but the number of columns will vary from row to row. Thus, we cannot do a simple fill like the one above. The best we could do is something like the following.

```
myimage(100, 70) = 255;
myimage(101, 69:71) = 255;
myimage(102, 68:72) = 255;
myimage(103, 69:71) = 255;
myimage(104, 70) = 255;
```

Changing this for larger diamonds will be quite a task. Instead, we need a better way to specify the shape, with a function to do the fill for us. The idea will be the same, however. For each row of the diamond, we set each column within a range to the fill value. The range will depend on the outline of the diamond; that is, we will not require that the diamond have a perfectly square shape. In fact, we will let most any shape defined by four corner points to be filled, for our purpose here.

Our diamond fill function will need a few pieces of information. First, we should specify an image (matrix) on which to draw the diamond shape. Also, we need the fill value to use. This will make our function flexible, and ultimately we will be able to draw a diamond of any color. Finally, we require a list of four corner points, to specify the diamond.

Since we are defining our requirements at this stage, we can call the parameters anything we like. We will use the following line to name the function, the inputs and the output.

```
function [myimage] = fill_diamond(myimage, fill_value, ...
    fourpoints)
```

The variable **fourpoints** specifies the four corner points. We expect it to have four rows and two columns, so that every row specifies an *x* and a *y* coordinate. This might be a bit complicated for the user. We help make this clear by adding the following lines to the header comments.

```
% Example:
%   myimage2 = fill_diamond(myimage, 255, ...
%      [x1, y1; x2, y2; x3, y3; x4, y4]);
```

Still, the user may find this confusing. We can further help by checking the number of rows and columns, and returning with an error if they do not meet our expectations.

```
[fp_rows, fp_cols] = size(fourpoints);
if ((fp_rows ~= 4) || (fp_cols ~= 2))
    str1 = 'fill_diamond expects four points,';
    str = strcat(str1, ' given as a 4x2 matrix.');
    error(str);
    return
end
```

The first line above finds the dimensions of variable **fourpoints**. Then we compare the number of rows and columns to 4 and 2, respectively, using the "not equal to" comparison, "~=" in MATLAB syntax. If either condition is not met, we print the error message and return.

Once we know that the dimensions are correct, we split the points into coordinates, according to the column. We call these coordinates **x** and **y**.

```
x = fourpoints(:,1).';
y = fourpoints(:,2).';
```

Yes, we could have made it work with the original **fourpoints** variable only, though this way should be more readable.

We might assume that the user will call our function with the four points in order from top to bottom. However, it makes for a user-friendly function if we allow the points to be in any order. We can sort them within the function. The top and bottom points are easy enough to discern. We simply choose the points with the minimum and maximum **y** coordinate, respectively, then swap these values (both **x** and **y** coordinates) with the first and fourth points. Next, we need to decide which of the remaining coordinates correspond to the diamond's left and right points. We define the left corner point as the one with the minimum x value.

```
[value, index] = min(x(2:3));
```

The variable **index** tells us what we really need, except for one slight problem. The selection **x(2:3)** forms a sub-array with two values, and when we locate the minimum value it will have an index relative to this sub-array. Thus, **index** will always have a value of 1 or 2, when we want it to tell us the index for **x**. We solve this by adding one to the index.

```
index = index + 1;
```

Now we need to take the value corresponding to **index** and swap it with the second values for **x** and **y**. Rather than first do the test to see if we need to make the swap, we will simply process it. The swap for the **x** values appears below.

```
tempx = x(2);
x(2) = x(index);
x(index) = tempx;
```

When no swap needs to be made, we will still execute these lines. Imagine what happens when **index** has a value of 2: we assign its value to a temporary variable, then assign **x(2)** to itself, then assign it to itself again. This takes a bit of time, but does not cause a problem.

In a similar fashion, we swap the **y** values.

```
tempy = y(2);
y(2) = y(index);
y(index) = tempy;
```

We use the same idea as before, simply changing the variable names.

For a given row, what columns should have their pixels filled? To answer this, consider the information that we have so far. The user specified a diamond shape, with four points defining the top, left, right, and bottom corners. The top and left corners form a line segment on the left side, as do the left and bottom corners. Refer to Fig. 3.12, which shows the four line segments that we interested in: top-left (connecting points 1 and 2, labeled segment **12**), top-right (**13**), bottom-left (**24**), and bottom-right (**34**).

 Given two points that specify a line, we need to know the column (**y** coordinate) to complete a point on the line, given the row (**x** coordinate). For normal cases, we just use the line equation. We also need to have some handling of extreme cases, such as when the **x** coordinate does not exist on the

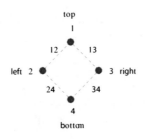

FIGURE 3.12   A typical diamond shape to fill.

sense to find the **y** coordinate when **x** has the value 3. We call our function
col_on_line.m, available on the CD-ROM. For balance, we also include
a program called **row_on_line.m**. As the name suggests, it returns the row
when given the column and two points defining a line. Below, we show how
an example call to the **col_on_line** function works. In this case, it finds
the line connecting points 1 and 2, then plugs in **row** to find the matching
column.

```
% given the row, find the column on line 1:2
col1 = col_on_line(row, x(1), y(1), x(2), y(2));
```

We will repeat the above call for the other lines, as shown below.

```
col2 = col_on_line(row, x(1), y(1), x(3), y(3));
col3 = col_on_line(row, x(4), y(4), x(2), y(2));
col4 = col_on_line(row, x(4), y(4), x(3), y(3));
```

By inspection, you should be able to tell that **col2** goes with line **13**, that
col3 goes with line **24**, and **col4** goes with line **34**. Notice that the order
of the points does not matter, that we can list either point 4 or point 3 first.
    We want to draw a horizontal line across the diamond shape, but as we see
in Fig. 3.12, there are two lines on either side that define our boundary. We
do not know when to stop using line segment **12** and switch to line segment
24 to get our left-most column. Therefore, we will get columns from both
line segments, then choose the right-most one of them. Fig. 3.13 shows why
we choose the right-most point of the two. In this figure, we see line segments
12 and **24** extended to the left. If we pick an arbitrary **y** coordinate, we see
the two possibilities, shown by the circle outlines. Obviously, the right-most
(maximum) one occurs on the diamond shape, so we choose it.

```
left_col = max(col1, col3);
right_col = min(col2, col4);
```

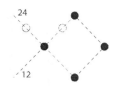

FIGURE 3.13   **The lines on the left give us two possibilities for a column coordinate.**

Similarly, we choose the left-most (minimum) of the two possibilities for the right column.

Now that we have the appropriate columns, we can fill the image for that row.

```
myimage(row, left_col:right_col) = fill_value;
```

Notice how much the above line compares to the original rectangle fill.

There are only a few details left to consider. First, we want the row to vary from the top-most row ($y(1)$) to the bottom-most row ($y(4)$). We create a loop, as below.

```
for row = y(1):y(4)
    % find columns
    % choose correct ones
    % fill the line
end
```

Inside this loop, we place the commands for finding the columns, choosing the correct ones, and filling the line. We saw these commands above, now we copy them into place.

We also must consider a detail about image colors. So far, we have only considered grayscale images. When we work with color images, we have three values for the color. We need to check for this case and alter all three color values, which appear as the third dimension of the image data. For example, instead of a fill being a single command:

```
myimage(row, left_col:right_col) = fill_value;
```

we would do a fill as three commands, one for each color, as shown below.

```
myimage(row, left_col:right_col, 1) = fill_value(1);
myimage(row, left_col:right_col, 2) = fill_value(2);
myimage(row, left_col:right_col, 3) = fill_value(3);
```

To decide which of the two possibilities our function should do, we can examine the third output from the **size** function, and act accordingly.

FIGURE 3.14 **Black diamond shape on a white background.**

 Program `fill_diamond.m` (on the CD-ROM) allows us to fill a diamond shape. We can test it with the following code.

```
myimage = 255*ones(128, 128);
myimage = fill_diamond(myimage, 0, ...
    [64, 32; 32, 64; 96, 64; 64, 96]);
imshow(myimage)
```

When we run the test code above, we get the image as seen in Fig. 3.14. Here, we have chosen a black diamond shape on a white background. See if you can alter this to make a white diamond on a black background.

Finally, we should note that the `fill_diamond` function does not work for rectangles. When we need to fill a rectangular area, we should use the code presented at the beginning of this section. Below, we have an example of this.

```
myimage = 255*ones(128, 128);
myimage(100:120, 50:90) = 0;
imshow(myimage)
```

The code above generates the black rectangle on a white background, as seen in Fig. 3.15.

FIGURE 3.15 **Black rectangle on a white background.**

3.15 DRAWING AN ENTIRE CUBE

Once we can fill a diamond shape, we can put many diamonds together to make our overall image. All that we do here is repeat the diamond drawing command for each set of defining points. Looking at the vertices of the frame in Fig. 3.11, we can see the defining points for the nine diamonds on each of the three faces. We also need to know what color to make each diamond. Up to this point, we have not dealt with the internal representation of the cube, but we address this now.

There are different ways that we could represent a cube with three rows and three columns of colors on each face. We will use a straight-forward approach that should be simple. Each face will be represented as a 3×3 grid of colors. The advantage here is that we can store each face as a matrix according to the color in the middle, and combine all six faces into a $3 \times 3 \times 6$ matrix. The disadvantage is that we must remember explicitly how the squares on different faces are related. That is, if we turn the cube to see a new face, the program must know how the new face should appear: which edge is closest to the viewer, and which row or column corresponds to it?

If we are drawing a physical object, the computer has no idea of connections. Suppose that we alter the corner where the three sides meet, perhaps blending the three colors. We would have to change three matrices, and it would not be obvious which entries would be altered. That is, we define that corner point by element $(3,3)$ of the top side, element $(1,3)$ of the left side, and element $(1,1)$ of the right side. We must keep track of the relationships ourselves.

You may have once mixed paints in an art class. If you take primary colors and mix them, like equal amounts yellow and red paint, you get another color (orange). By adding more yellow, you get a brighter orange. For us, we specify the red, green, and blue components, the primary colors for a computer display, to make the desired color. The `colors` matrix, encodes this color information. We define the first seven colors as follows. The actual definition lists more (16) different colors, but these are enough to get started.

```
colors = [255, 255, 255;   % 1 white
          255, 255,   0;   % 2 yellow
          255,  90,  40;   % 3 orange
            0,   0, 255;   % 4 blue
            0, 128,   0;   % 5 green
          128,   0,   0;   % 6 red
            0,   0,   0];  % 7 black
```

The colors are chosen according to how they appear. For example, if you draw a rectangle on an image with 255 for the first color (red component), 90 for the second color (green component), and 40 for the third color (blue component), then the rectangle should appear as orange. These values are a matter of personal preference; you could alter them to make a brighter green or a deeper blue color.

White will correspond to the first face, then yellow, orange, etc. This order is arbitrary, but we must be consistent whatever order we use. To make this work, we assign each color name a value. This way, we can refer to yellow as the variable `yellow`, instead of the less natural designation **2**.

```
white = 1;
yellow = 2;
orange = 3;
blue = 4;
green = 5;
red = 6;
black = 7;
```

There are actually more colors defined in the `cube.m` file, but these are enough to convey the idea.

 Next, we populate each of the faces with colors. This can be a bit tedious, depending on how we want it to look. Here, we use the same simple state as in the paper cube photograph from the CD-ROM.

```
redface = [white, white, white;
           white, red,   white;
           white, white, white];
```

The red face provides a nice example. Red and orange faces are oriented such that the cube is resting on the white face. The top row corresponds to the squares bordering the green side. Note that the orientation is an arbitrary choice, but a choice that we must make. For other sides, we orient them and number the squares on their faces differently. The green face may be visible when the orange and red faces can be seen. We will number the green face's squares with the bottom row bordering the orange face, and the third column bordering the red face.

For the yellow and blue faces, we orient them such that the cube rests on the green face, which would be the case when we flip the cube around. The top rows correspond to the squares bordering with the white side. For the white side, we number the squares as we did the green side, with the bottom row bordering the blue face, and the third column bordering the yellow face.

To hold all of this information, we create a three-dimensional matrix. Each face has three rows and three columns, and there are six faces total. So we create a new matrix, **mycube**, with the third dimension corresponding to the face. Below is an example of how we store one color's side.

```
mycube(:,:,blue)   = [blueface];
```

With our example cube, each face has a different color in the center square, with white squares around it.

Showing the cube involves a few steps, ones that we have already seen. First, we create a blank image along with variables that we need, like image dimensions.

```
MaxRow =    700;
MaxCol =    700;
im2 = 255*ones(MaxRow, MaxCol, 3);
```

Next, we specify the vertex points for our cube, just as we did to draw the frame. We also specify the left side and the right side. Since we already showed the top face's points, we include the left side vertices below. These go with whatever face will appear on the lower-left.

```
y2 = [y1(7), y1(11),  y1(14),  y1(16), ...
        249,     297,    356,      427, ...
        376,     431,    496,      572, ...
        471,     533,    601,      682];

x2 = [ x1(7), x1(11),  x1(14),  x1(16), ...
         42,     122,     221,   x1(16), ...
         51,     132,     227,   x1(16), ...
         57,     140,     231,   x1(16)];
```

So far, we are not doing any different from drawing the frame. We might be tempted to draw the lines next, but they will need to appear over top of the squares. Drawing them now means that they will be covered up by the colors. Therefore, we fill in the colors next. In the code below, we have an example square being colored. Since the square appears to us at an angle, we use the **fill_diamond** function.

```
leftface = orange;

[im2] = fill_diamond(im2, colors(mycube(1,1, leftface),:), ...
    [x2(1), y2(1); x2(5), y2(5); x2(2), y2(2); x2(6), y2(6)]);
```

The first parameter gives the current image; when the function call ends, the updated image will be returned and stored in the same matrix. The second parameter looks difficult, but it specifies a color. To understand it, we examine it in one more level of detail. Variable `mycube` holds the state of the cube, containing information about what colors go where. Which face appears on the left depends on what sides we want to view, so we set up a variable `leftface` to keep track of this. Remember that `orange` simply holds a number, 3 in this case. Thus, `mycube(1,1, leftface)` returns the element in row 1, column 1, on the orange side. That value will simply be a number between 1 and 6, indicating the color. For the values that we used, this maps to 3, the color for orange. After resolving this, we have the code `colors(3,:)`, which gives us all columns on row 3 of the `colors` matrix, which in turn specifies the red, green, and blue component to make the desired color. The other parameter to the `fill_diamond` function specifies the four points that define the diamond shape. We repeat the call to `fill_diamond` over and over again, until we have drawn all 27 of the squares.

Next, we draw the lines. First, we find the points on the line, then we draw each point on the image.

```
[newx1, newy1] = points_on_line(x2(1), y2(1), x2(4), y2(4));
for k=1:length(newy1)
    im2(newy1(k), newx1(k), :) = 0;
    im2(newy1(k)+1, newx1(k)+1, :) = 0;
end
```

Color value 0 corresponds to black. As before, we actually draw two lines side by side to make them stand out. We need to repeat this for each line that we used in the frame drawing function. Finally, we show the image.

```
image(uint8(im2));
```

The conversion to `uint8` turns the color values of type `double` into the 8-bit values that the `image` function expects.

We can show the whole cube by repeating the above code, as needed. However, we can only show three sides at a time, just as we can only view three sides of a real cube at a time. The next section covers how we can view other sides of the cube besides the default three sides.

3.16 ADJUSTING OUR VIEW

Holding a cube, we can only see three sides at a time. When we change our orientation, we view another three sides. With a physical cube, we simply turn it over in hand. With a virtual cube, we need to specify different sides to view. We will not worry about animation or gradually changing the view.

The trick here is to have a default view, then change it depending on the desired view. We will always show three faces, but where we locate the faces may be different. Consider the top face from the mesh cube in Fig. 3.11. We want to keep the perspective, where the bottom corner appears closer to the observer than the other corners. Imagine rotating the whole frame around the point where all three faces meet, until the top face becomes a bottom face. This is how the cube should look when flipped. While the concept is easy, no simple option exists to do this.

As with most programming problems, there are multiple ways to solve this. We could possibly subtract the coordinates of the corner where all three faces meet from all x and y coordinates, to make that point an origin. Then we could transform each point to polar coordinates, add 180 degrees to the angle, then transform back and add the corner points' original values. Actually, if we did this, we would find that some points exist outside the matrix. That is, rotating around point (339, 271) means that other points would have negative coordinates, since this point does not align with the center of the frame. Some points would have negative coordinates, which is no problem for a graph in general, but this does not work when the coordinates map to matrix indices. In short, we would need to correct the coordinates by adding a constant.

However, we will take a different route that involves fewer calculations. We treat each visible side separately. For the top side, rotate it around a point and figuratively slide it down the image to a new location. The other sides will be rotated and adjusted in a similar way. In other words, we will draw the "left face" on the lower left when the flip variable has an even value, but put it in the upper right otherwise.

To clarify this, see Fig. 3.16. It shows how we number the vertex points from top to bottom and left to right. In the figure, we have numbers for the faces to make them non-location specific.

Now when we take the flip into account, we keep the shapes of each face. Face 1 appears on the bottom, while Face 2 goes from the lower left to the upper right. Face 3 goes from the lower right to upper left. Fig. 3.17 shows this. Try to envision the frame of each face rotating and sliding from one position (Fig. 3.16) to another (Fig. 3.17). One advantage of this approach is that the points remain in the same relation to each other. Drawing the line between points 5 and 8 on the first frame still must be done on the flipped frame. Another, more important, advantage is that all point coordinates can be easily rotated. In other words, we do simple operations (additions,

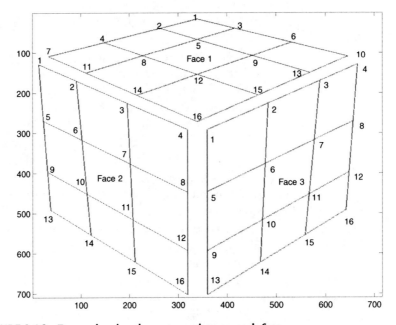

FIGURE 3.16   Frame showing the vertex points on each face.

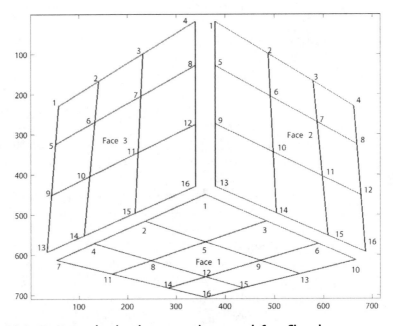

FIGURE 3.17   Frame showing the vertex points on each face, flipped.

subtractions, and re-arranging values) to map the points on the original frame to their respective locations on the flipped frame.

You may have noticed that some of the lines overlap. For example, the line segment from points 7 and 16 on Face 1, with the original orientation, will be the same as the line segment from points 1 and 4 on Face 2. To see this, examine Fig. 3.16, and imagine that the three faces are brought together until they line up. While this does not take much processing time, we can eliminate the extra line drawing by enclosing the code to draw the three bordering lines within an **if** statement. As you can see from the flipped version, we have three other, different lines that will overlap. These can be drawn as needed with an **else** clause. The pseudo-code below shows this for one example.

```
if we have a flip,
    draw a line between Face 3's points 1 to 4
else
    draw a line between Face 3's points 13 to 16
end
```

One of those two lines will be overlapped, depending on the **flip** value.

Now that we have a basic idea about how the points are laid out, and how the code draws lines between them, we can get to how, specifically, we move the faces around. For Face 1, we need to do two things: reverse the order of the points and move them to a new location. The diamond shape made by points 1, 2, 3, and 5 on Face 1 has the farthest distance from the observer, and appears the smallest. When we flip the shape around, though, the diamond made by points 12, 14, 15, and 16 is the farthest away (smallest). We can exchange the first points with the last points with a command like the following.

```
points = points(length(points): -1: 1);
```

This command reverses the order of the points in the array **points**.

Fig. 3.18 shows how the points on Face 1 will appear after the flip. Envisioning what happens to the points as a whole can be easier by considering a single point. In this figure, we see how the values for point 2's coordinates become point 15's coordinates after the flip. We use point 16 as a fixed point, and rotate the other points around it. First, we find the distance (subtraction) between point 16 and our example, point 2, and store the distance as **dx** and **dy**. Whatever the distance might be along the x-axis, we want to copy on the other side of the y-axis. In other words, however much point 2 is "to the left" of point 16, we want to make the new value that much "to the right."

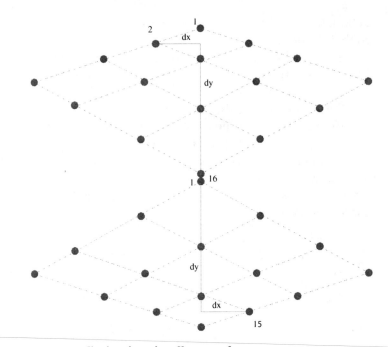

FIGURE 3.18   How flipping the cube affects one face.

Adding the **dx** value to point 16 achieves this aim. The only thing left to do is reverse the order of the new values, just as we did with the variable **points** above. This specifies the new *x* coordinate; the *y* coordinate can be found in a similar fashion, except for one small issue. We do not actually want Face 1 rotated in place around point 16, as Fig. 3.18 might suggest. We need to have room on our image for the other two faces, so it makes sense to slide the new version of Face 1 along the *y*-axis. Thus, we simply add a constant to the new *y*-coordinate before reversing the new points. The code segment below shows how to do this.

```
dx = x1(16) - x1;
dy = y1(16) - y1;
y1 = dy(length(dy):-1:1) + y1(16) + 158;
x1 = dx(length(dx):-1:1) + x1(16);
```

Variables **x1** and **y1** store the *x* and *y* coordinates for Face 1, respectively. You may wonder why the constant 158 was chosen. It was a rather arbitrary choice; but it works well with respect to the image size, and room needed for the other faces.

For the other two faces, we perform a similar rotation. For Face 2, we want to figuratively spin it 180 degrees around point 4, the equivalent to point 16 on the first face. After such a rotation, point 16 would be the highest point on Face 2, followed by point 15, etc. We renumber the points as we did above to make these points 1, 2, etc.

You may observe that the shape for Face 2 does not dramatically change; it looks as if we could just slide the face to the upper right. That approach creates a problem, though, since we lose the perspective. The diamond shape defined by points 3, 4, 7, and 8 on Face 2 makes the largest such shape on this face. It would look odd to the viewer for the largest square to also be the furthest away. It should be apparent why we need to take the rotation approach to keep the illusion of distance to the viewer.

 With these considerations, we can show a full-color, three-dimensional cube as an image in MATLAB. The reader is encouraged to run the programs included on the CD-ROM, especially `cube.m`. It is a "turn-key" type program; all the user needs to do is type `cube` in the command line interface. It calls `draw_cube.m`, which works to display the image. The variable `flip` is defined at the top of the `cube` program, and controls which sides are visible. Adding one to the `flip` value is like rotating the whole cube away from the viewer. The `flip` variable should have a value between 0 and 3. After a value of 3, an additional flip brings the cube back to the original orientation. Testing the program versus a physical cube reveals that all sides are shown correctly. You may want to alter `cube.m` to color code a cube of your choice.

3.17 EPILOGUE

This chapter of projects presents solutions in a finalized form. The programs were developed in an iterative fashion; one part at a time, with functionality added incrementally. What the chapter does not show are all the syntax errors, logic issues, and downright silly mistakes that went into making these programs. Rarely will a program work correctly the first time that you run it.

When things do not work, look at the line indicated by MATLAB. Looking over code for errors, often we see what we want to see: code that works. But when told that a particular line does not work, the experienced programmer can often spot the problem. Syntax errors, such as misspelling a function, leaving out a close parenthesis, or neglecting to include a terminating single quote, are easy to fix. Logic errors are often much harder to discover, let alone correct. These stem from the code specification failing to match up with the programmer's expectations. This can include bizarre results from a program that runs, such as a negative amount for a dollar figure, a calculated speed in the millions of meters per second for a falling object, or perhaps a numeric result when you expect a string. Sometimes the problem causes the program to run longer than you have the patience to wait. Is it an infinite loop, or just a slow program? To deal with logic errors, making modular functions helps. A function will have a limited number of inputs and calculate a limited number of outputs, even when the inputs or outputs consist of large matrices. By checking each function's results in turn, we should be able to see when the results differ from our expectations. When dealing with large matrices, print a subsection to the screen, e.g., `mymatrix(1:10, 1:10)`. Checking a corner of the values may reveal unexpected values. The minimum and maximum values also help to point out problems, e.g., `min(min(mymatrix))` and `max(max(mymatrix))`. We do this twice for a two-dimensional matrix, since performing the `min` or `max` function once returns a vector. For example, sometimes the minimum and maximum values are both zero indicating that the matrix was initialized but not properly updated, so checking these values comes in handy.

Writing a program is a learning experience. If you cannot see the problem, often someone else can. When something works well, add to it a bit at a time, and be sure to keep backup copies of your work. Keep comments in your code. Often you can write better code after explaining to yourself what you intend to do, especially when dealing with a complex problem. And when you realize that a problem is too complex, try to break it down into smaller problems. The more time you spend working with the computer, the better you will know how to make it do what you want. So try not to get discouraged when things do not work. Trial and error provide a great way to learn programming, and an interactive programming environment like MATLAB gives you a good platform for this.

4

SOLUTIONS TO
THE PROBLEMS

S olutions to the problems formulated in this book are given in this
chapter. Readers should work out these problems on a computer
as they read the book. Most of the solutions are also given in the
form of MATLAB m-files or Simulink systems in the software accompanying this book. The names of the appropriate m- or mdl-files are given
with each solution. An indication of which call is to be used for the solutions in MATLAB can (in the case of m-files) be read out by entering `help
<Filemame_without_ending>`. The relevant comments for Simulink
systems appear in the system window.

You should make sure that the installed MATLAB search path (see
Section 1.2.9) contains the folder with the files from the accompanying
software.

4.1   SOLUTIONS TO THE MATLAB PROBLEMS

Solution to Problem 1 (file: soldefmat.m)

```
% Define the matrix M in MATLAB

M = [ 1   0   0;...
      0   j   1;...
      j  j+1 -3]


% Define the number k in MATLAB

k = 2.75


% Define the column vector v in MATLAB
```

```
v = [1; 3; -7; -0.5]


% Define the row vector w in MATLAB

w = [1, -5.5, -1.7, -1.5, 3, -10.7]


% Define the row vector y in MATLAB

y = (1:0.5:100.5)
```

Solution to Problem 2 (file: solmatexp.m)

```
% Define the matrix M in MATLAB

M = [ 1   0   0;...
      0   j   1;...
      j  j+1 -3];


% Define the matrix V using M

V = [M M; M M]
```

Obviously it would be very tedious to define large matrices in this way by duplication. Thus, the MATLAB command **repmat** is available for this task. The above matrix can be defined using this command as follows:

```
V = repmat(M,2,2)
```

The solutions for parts (2) to (4) of this problem are as follows:

```
% Erasing row 2 and column 3 of V
% and saving as V23
V23 = V;              % Save V, otherwise the matrix
V23(2,:) = [];        % would be erased by these operations
V23(:,3) = [];


% Define the vector z4 with the number of rows of V (here 6)

z4 = [1 2 3 4 5 6];


% Attach the row vector z4 to row 4 of V

V(4,:) = z4;
```

```
% Change the term 4,2 of matrix V

V(4,2) = j+5


V =

  Columns 1 through 3

     1.0000                      0                    0
          0                      0 + 1.0000i    1.0000
          0 + 1.0000i     1.0000 + 1.0000i   -3.0000
     1.0000               5.0000 + 1.0000i    3.0000
          0                      0 + 1.0000i    1.0000
          0 + 1.0000i     1.0000 + 1.0000i   -3.0000

  Columns 4 through 6

     1.0000                      0                    0
          0                      0 + 1.0000i    1.0000
          0 + 1.0000i     1.0000 + 1.0000i   -3.0000
     4.0000               5.0000                6.0000
          0                      0 + 1.0000i    1.0000
          0 + 1.0000i     1.0000 + 1.0000i   -3.0000
```

Solution to Problem 3 (file: none)

The matrix **N** is constructed by writing the column vector corresponding to $\vec{r}$ six times in succession in a matrix bracket:

```
>> rs = [j; j+1; j-7; j+1; -3];
>> N = [rs, rs, rs, rs, rs, rs];
```

Solution to Problem 4 (file: none)

The row vector cannot be attached, since it only has five components, while the matrix **N** has six columns.

Solution to Problem 5 (file: none)

First, erase all the defined variables using

```
    clear                   % alternative:  clear all
```

The command

```
    z4 = [1 2 3 4 5 6];
```

can, for example, be reconstructed by repeatedly pressing the ↑ and ↓ keys. Here, it is understood that you have already entered this command into the command window in Problem 2. If you have created the command by entering **soldefmat**, then the above command has not, of course, been saved. In that case it has to be entered again.

If the beginning of the command to be reconstructed is known, then the search can be shortened by entering the starting characters (as uniquely as possible) of the command before pressing the ↑ and ↓ keys. As an example, for the above vector **z4**, this would be

```
z4 =
```

If the command-history window is open, then a double-click on the row

```
z4 = [1 2 3 4 5 6];
```

is sufficient for the reconstruction and the command will again be executed in the command window.

Solution to Problem 6 (file: none)

If it is not already open, open up the workspace browser and enter a check mark in the *Desktop* menu next to *Workspace*. Then double-click on the symbol for V. In that way you get the Array Editor window shown in Fig. 4.1.

Select the fields in row 5 in succession and type in 0.
Then a subsequent MATLAB call for V gives

```
>> V
```

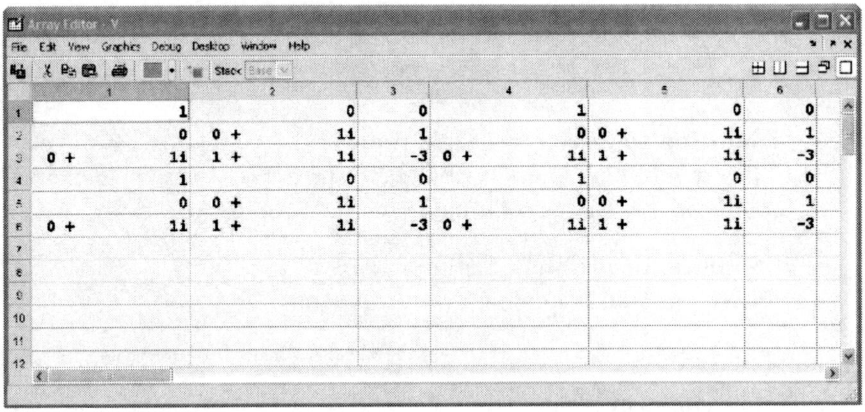

FIGURE 4.1 The Array Editor with the matrix V before filling row 5 with zeroes.

```
V =

  Columns 1 through 3

   1.0000                     0                      0
        0                     0 + 1.0000i   1.0000
        0 + 1.0000i    1.0000 + 1.0000i  -3.0000
   1.0000               5.0000 + 1.0000i   3.0000
        0                     0                      0
        0 + 1.0000i    1.0000 + 1.0000i  -3.0000

  Columns 4 through 6

   1.0000                     0                      0
        0                     0 + 1.0000i   1.0000
        0 + 1.0000i    1.0000 + 1.0000i  -3.0000
   4.0000               5.0000             6.0000
        0                     0                      0
        0 + 1.0000i    1.0000 + 1.0000i  -3.0000
```

This shows that the row of zeroes has been entered.

Solution to Problem 7 (file: none)

In the following we describe one of the many ways data can be imported.

Once you have opened the Excel file `ExcelDatEx.xls` in the accompanying software, select the data block by clicking in the cell **A1** and (holding the **shift** key down) **C15**. The data block is copied into the Windows interface using the **Edit - Copy** command.

Next, select the symbol **New Variable** from the icon toolbar in the **Workspace** window. In the name field which then appears for the variable, enter **TheExcelData**. A variable with this name will be set up with value 0. Then open the Array Editor by clicking the **Open Selection** icon. In the open Array Editor select the menu command **Edit - Paste**. The data will immediately be inserted and are now available in the workspace, as the following example shows:

```
>> whos
  Name                  Size            Bytes  Class

  TheExcelData          15x3              360  double array

Grand total is 45 elements using 360 bytes
```

Solution to Problem 8 (file: aritdemo.m)

Enter the command **aritdemo** in the MATLAB command window and follow the commands on the screen.

Solution to Problem 9 (file: solmatops.m)

The standard scalar product of two vectors $\vec{x}$ and $\vec{y}$ $\mathbb{R}^n$ is defined as

$$\langle \vec{x}, \vec{y} \rangle = \sum_{k=1}^{n} x_k \cdot y_k \, .$$

This can be set up under MATLAB as follows:

```
% Define the vector x

x = [1 2 1/2 -3 -1];

% Define the vector y as a COLUMN vector

y = [2; 0; -3; 1/3; 2];

% Calculating the scalar product <x,y> with a
% matrix operation

skp1 =  x*y

% Calculating the scalar product <x,y> with a
% field operation
                        % Define the vector y as a ROW vector
y = [2, 0, -3, 1/3, 2];

fop =  x.*y;       % contains the vector of the product
                   % of the components
skp2 =  sum(fop)   % Takes the sum of these components
                   % with the MATLAB function sum
```

The solutions for parts (2) and (3) of this problem look like:

```
% The product of the matrices A and B

A = [-1 3.5 2;  0 1 -1.3;  1.1 2 1.9];
B = [1 0 -1;  -1.5 1.5 -3;  1 1 1];

A*B

% Creating a diagonal matrix with field operations
```

```
E3 = [1 0 0;  0 1 0;  0 0 1];  % a 3x3 unit matrix

C = A.*E3                      % field operation
```

A unit matrix can also be defined using the MATLAB command **eye**. All that has to be specified is the size of the matrix; in the above case, this would be

```
E3 = eye(3);
```

Solution to Problem 10 (file: solleftdiv.m)

```
% Define the matrix A

A = [ 2 2;...
      1 1];

% Define the vector b

b = [2; 1];


% Perform left division

x = A\b

Warning: Matrix is singular to working precision.
> In solleftdiv at 13

x =

    NaN
    NaN
```

As explained in Section 1.2.2, $A\backslash\vec{b}$ must be interpreted as $A^{-1}\vec{b}$. But there is no inverse matrix for **A**. This matrix is referred to as *singular*. This explains the warning message. The components **NaN** ("not a number") are assigned to the result vector.

Solution to Problem 11 (file: solrightdiv.m)

```
% Define the matrix A

A = [ 2 2;...
      1 1];

% Define the vector b
```

```
b= [2 1];

% Perform right division

x = b/A

Warning: Matrix is singular to working precision.
> In solrightdiv at 12

x =

   Inf    -Inf
```

$\vec{b}/A$ must be interpreted as $\vec{b}A^{-1}$. The reaction of MATLAB is explained as in the preceding solution. The values Inf (∞) and $-\text{Inf}$ ($-\infty$) are assigned to the result vector.

Solution to Problem 12 (file: none)

The matrix M can be inverted using *left* division by the *unit matrix*:

```
>> M = [1 1 1 ; 1 0 1; -1 0 0];
>> Minv = M\eye(3)

Minv =

    0     0    -1
    1    -1     0
    0     1     1
```

The last instruction corresponds to a realization of the equation $Minv = M^{-1}E_3$. Since $M^{-1}E_3 = M^{-1}$, Minv contains the desired inverse matrix.

Solution to Problem 13 (file: sollogops.m)

```
% Define the matrices A and B

A = [1 -3 ;0  0];
B = [0  5 ;0  1];

% Determine logical OR

resOr = A|B;
```

This yields the result

```
>> resOr
```

```
resOr =

     1      1
     0      1
```

Only the (2, 1) term is equal to 0 (logically false) for the two matrices A and B. For all the other terms, there is at least one number that is interpreted as logically true. Thus, the result of the **OR** for these terms is always 1 (true).

```
% Determine the logical negation of A

NegA = ˜A;


% Determine the logical negation of B

NegB = ˜B;
```

This yields the result

```
>> NegA

NegA =

     0      0
     1      1

>> NegB

NegB =

     1      0
     1      0
```

A term that is $\neq 0$ (true) is converted into 0 (false) and a term that is 0 (false) is converted into 1 (true).

```
% Determine xor of A and B

resXor = xor(A,B);
```

The result is

```
>> resXor

resXor =
```

```
    1      0
    0      1
```

The result is 1 if the numbers in a given term of *A* and *B* are *different* when interpreted as *logical values*, otherwise the result is 0.

Solution to Problem 14 (file: solrelops.m)
The MATLAB instructions

```
% Define the vectors x and y

x = [1   -3    3   14   -10    12];
y = [12   6    0   -1   -10     2];

% The less than and less than or equal operators

resultLess   = x<y;
resultLessEq = x<=y;

% greater than or equal operator, and

resultGreaterEq = x>=y;

% equal and unequal operators

resultEq = x==y;
resultNotEq = x~=y;
```

yield the following results when applied to the vectors **x** and **y**:

```
>> resultLess
resultLess =

     1     1     0     0     0     0

>> resultLessEq

resultLessEq =
     1     1     0     0     1     0

>> resultGreaterEq

resultGreaterEq =

     0     0     1     1     1     1
```

```
>> resultEq

resultEq =

     0     0     0     0     1     0

>> resultNotEq

resultNotEq =

     1     1     1     1     0     1
```

Solution to Problem 15 (file: sollimvals.m)

```
% Define the matrix C

C = [  1  2   3   4  10; ...      % with ... a command line
     -22  1  11 -12   4; ...      % can be broken
       8  1   6 -11   5; ...
      18  1  11   6   4 ]

% Define the comparison matrices

U = -10*ones(4,5);
O =  10*ones(4,5);

% Make the comparisons

X = C>=U;                         % 1, if the term in C >=-10
Y = C<=O;                         % 1, if the term in C <= 10

% Positions of the numbers which lie between -10 and 10

P = X&Y;
```

First, two matrices with the same dimensions as C are defined whose terms are only -10 or 10. Then C is compared with these matrices. The matrix P contains a 1 if the terms of C lie within these limits, otherwise a 0.

```
% Numbers lying outside the interval [-10,10], set to 0

result = C.*P
```

Here, *term-by-term* multiplication of *P* by *C* produces the desired result. The values of *P* are again interpreted as numbers, rather than logical values.

Executing this sequence of instructions by entering the command `sollimvals` yields the following:

```
>> sollimvals

C =

        1     2     3     4    10
      -22     1    11   -12     4
        8     1     6   -11     5
       18     1    11     6     4


result =

        1     2     3     4    10
        0     1     0     0     4
        8     1     6     0     5
        0     1     0     6     4
```

Finally, it should be noted that the comparisons

```
X = C>=U;                    % 1, if the term in C >=-10
Y = C<=0;                    % 1, if the term in C <= 10
```

can also be replaced by the instructions

```
X = C>=-10;                  % 1, if the term in C >=-10
Y = C<=10;                   % 1, if the term in C <= 10
```

in the present case. Of course, matrices (arrays) that are connected by relational operators have the *same dimensions*, although a connection *with a number* is an exception. In that case, the number is, as desired, connected with every component of the field.

Solution to Problem 16 (file: soldiagselect.m)

```
% Define the matrix D

D = [  7  2   3  10; ...      % with ... a command line
      -2 -3  11   4; ...      % can be broken
       8  1   6   5; ...
      18  1  11   4 ]
```

```
% Define the unit matrix
% Otherwise (and better), the command
% E = eye(size(D)); can be used

E = [  1  0   0    0; ...
       0  1   0    0; ...
       0  0   1    0; ...
       0  0   0    1];

% create the logical unit matrix out of this

Elog = logical(E)

% select the diagonal

theDiag = D(Elog)
```

The unit matrix is converted by the function **logical** into a logical array with which the matrix **D** can be indexed. The result is a *column vector* containing the diagonal.

This is illustrated by executing the above sequence of instructions, which can be started by entering the command **soldiagselect**:

```
>> soldiagselect

D =

       7      2      3     10
      -2     -3     11      4
       8      1      6      5
      18      1     11      4

Elog =

       1      0      0      0
       0      1      0      0
       0      0      1      0
       0      0      0      1

theDiag =

       7
      -3
```

```
        6
        4

>> whos
    Name            Size                    Bytes  Class

    D               4x4                       128  double array
    E               4x4                       128  double array
    Elog            4x4                        16  logical array
    theDiag         4x1                        32  double array

Grand total is 52 elements using 304 bytes
```

The concluding call of whos again shows that Elog is a logical field.

Solution to Problem 17 (file: solfuncdef1.m)

```
% Define time vector

t=(0:0.1:10);


% Define signal value

s = sin(2*pi*5*t).*cos(2*pi*3*t) + exp(-0.1*t)
```

A couple of comments should be made about this solution. The definition

```
% Define signal value

s(t) = sin(2*pi*5*t).*cos(2*pi*3*t) + exp(-0.1*t)
```

is a common error. MATLAB responds to it with the warning

```
??? Subscript indices must either be real
positive integers or logicals.
```

Unlike in the customary mathematical notation, function values *cannot* be defined in MATLAB in the form $s(t)$. MATLAB takes s(t) to mean that a term with an index from t is to be chosen from the vector s, but in the above example objected that the number of terms in the set of indices is not integral. The rest of the definition is ignored entirely.

The correct definition of the value vector automatically assigns the corresponding functional value to each value of t.

In addition, when defining function values it is important to keep in mind that the multiplication

```
sin(2*pi*5*t).*cos(2*pi*3*t)
```

involves a *field operation*. The instruction

```
sin(2*pi*5*t)*cos(2*pi*3*t)
```

would mean that here the value vector **sin(2*pi*5*t)** is to be combined with the value vector **cos(2*pi*3*t)** according to matrix multiplication. Because of the dimensions of the vectors (both are $n \times 1$ vectors), this is not defined, so MATLAB responds with

```
>> sin(2*pi*5*t)*cos(2*pi*3*t)
??? Error using ==> mtimes
Inner matrix dimensions must agree.
```

Solution to Problem 18 (file: solfuncdef2.m)

```
% Define the time vector

t = (0:0.1:10);

% Define signal values

s = sin(2*pi*5.3*t).*sin(2*pi*5.3*t)

% Alternatively, this can be defined as

s2 = sin(2*pi*5.3*t).^2
```

Note that, here as well, multiplication and the taking of a power (squaring) are meant to be done *term-by-term*, and must accordingly be treated as field operations.

Solution to Problem 19 (file: solround1.m)

```
% Define the time vector

t = (0:0.1:10);

% Define signal values

s = 20*sin(2*pi*5.3*t);

% Round s toward infinity (up) using the function ceil

s2infty = ceil(s)

% Round s toward 0 with the function fix
```

```
s2zero = fix(s)
```

The rounding functions `ceil`, `fix`, `round`, and `floor` can be used to round numbers off to integers in different ways. These functions differ only in the *direction* of rounding. For example, 1.25 rounds down (toward 0) to 1 and up (toward ∞) to 2, but -1.25 rounds toward 0 to -1 and toward ∞ also to -1.

Here is the result of the above calculations in MATLAB for the first six values:

```
>> s(1:6)

ans =

  Columns 1 through 5

        0   -3.7476    7.3625  -10.7165   13.6909

  Column 6

  -16.1803

>> s2infty(1:6)

ans =

     0    -3     8   -10    14   -16

>> s2zero(1:6)

ans =

     0    -3     7   -10    13   -16
```

Solution to Problem 20 (file: solround2.m)

```
% Define the time vector
t=(0:0.1:10);

% Define signal values

s = 20*sin(2*pi*5*t);

% Round s toward the next integer using the function round
```

```
s2round = round(s);

% Output of the first 6 values of s and s2round into
% a matrix with two rows

[s(1:6); s2round(1:6)]


ans =

  1.0e-013 *

        0    0.0245   -0.0490   -0.2818   -0.0980    0.1225
        0         0         0         0         0         0
```

All the values of $s(t)$ have evidently been rounded toward 0. A more precise look at the first row shows that the values of **s** all are on the order of 10^{-13} (i.e., very close to 0). If the theoretically expected values are examined more closely, we find that the sine of all multiples of $0.1 = \frac{1}{10}$ is

$$ s\left(k\frac{1}{10}\right) = 20\sin\left(2\pi\,5k/10\right) = 20\sin\left(k\pi\right) = 0 \quad \text{for all } k \in \mathbb{N}. $$

The theoretical values are also all 0. The values we see in MATLAB's response are within the limits of error of the computations.

We shall return to this phenomenon again when we discuss the graphical capabilities of MATLAB. In this case, a graphical display can lead to strange results and erroneous interpretations.

Solution to Problem 21 (file: sollogarit.m)

```
%  Define the vector

b = [1024 1000 100 2 1];

% Calculate the base 10 logarithm

log10ofb = log10(b)

% Calculate the base 2 logarithm

log2ofb = log2(b)
```

```
log10ofb =

    3.0103    3.0000    2.0000    0.3010         0


log2ofb =

   10.0000    9.9658    6.6439    1.0000         0
```

Solution to Problem 22 (file: none)

Entering the command **help cart2pol** in the command interface yields:

```
>> help cart2pol

 CART2POL Transform Cartesian to polar coordinates.
    [TH,R] = CART2POL(X,Y) transforms corresponding
    elements of data stored in Cartesian coordinates
    X,Y to polar coordinates (angle TH and radius R).
    The arrays X and Y must be the same size (or
    either can be scalar). TH is returned in radians.

    ...
```

This problem is then solved using the following MATLAB commands:

```
>> [th, r] = cart2pol(points(:,1),points(:,2))

th =

    1.1071
    0.6435
    0.7854
         0
    0.1107


r =

    2.2361
    5.0000
    1.4142
    4.0000
    9.0554
```

```
>> polarc2 = [r,th]

polarc2 =

    2.2361    1.1071
    5.0000    0.6435
    1.4142    0.7854
    4.0000         0
    9.0554    0.1107
```

Solution to Problem 23 (file: solfsequence.m)

Calling this sequence of commands (i.e., calling `solfsequence` via MATLAB) yields

```
>> solfsequence
??? Error using ==> plot
Vectors must be the same lengths.

Error in ==> solfsequence at 7
plot(t,[sinfct, cosfct, expfct])
```

The mistake thus lies in the plot command and is attributable to the different lengths of `t` and `[sinfct, cosfct, expfct]`. Although the vectors `sinfct`, `cosfct`, and `expfct` all have the same length as `t`, as can easily be confirmed by entering `whos`, the vector `[sinfct, cosfct, expfct]` is *three times as long*, since it is produced by writing the vectors with the functional values one after another (separated by commas).

The correct way is to write the vectors *one under the other* (separated by semicolons) in the form `[sinfct, cosfct, expfct]`, so that in the plot command `t` can be assigned to each of the row vectors in this $n \times 3$ matrix and the plot can be made correctly.

Solution to Problem 24 (file: solwhatplot.m)

Fig. 4.2 shows the MATLAB graph generated by this sequence of commands.

Only the exponential function is recognizable. The cosine shows up as a *triangular wave* and the sine as a *zero line*.

This result is attributable to a phenomenon mentioned previously in the solution to Problem 20. The choice of times in the vector `t` (here multiples of 0.5) for evaluating the sine function and the frequency of the sine wave (5 Hz) mean that the sine is evaluated at its *zeroes*. The cosine is evaluated

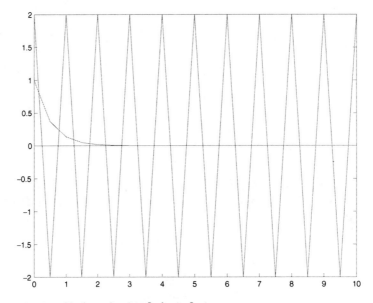

FIGURE 4.2 **Graphical result of** solwhatplot.

at the points

$$2\pi \cdot 3 \cdot 0.5k = 3k\pi, \ k \in \{0, 1, \cdots, 20\}.$$

There the cosine alternates between $+1$ and -1. Between these values, plot joins the values with a straight line, thereby producing the impression of a triangular wave, and in the case of the sine, a zero line. In terms of systems theory this phenomenon is related to the so-called *sampling theorem*, which effectively states how signals must be sampled in order to avoid losing information. In the above examples the premises of this theorem are violated, so information has been lost. We won't go into these theoretical aspects here. The interested reader is referred to the relevant specialist literature.

This example shows that caution is required in interpreting graphics. It is always best to use additional information for checking a graphical result and not to accept the result blindly.

Solution to Problem 25 (file: solplotstem.m)
Study the source text for the script file solplotstem and experiment with it as you make appropriate changes.

The original solution in solplotstem shows, in particular, that the command stem is extremely unsuitable for small step sizes and is preferable for sparse data sets. The opposite is true of the command plot.

Solution to Problem 26 (files: sollogplot1.m, sollogplot2.m)
For reasons of space, here we only reproduce the source text of
`sollogplot1`:

```
% Definition of the frequency vector omega

omega = (0.01:0.01:5);

% Definition of the value of the transfer function H(j*omega)
% Pay attention to field operations.

H1 = 1./(j*omega);

% Graphical display of the amplitude of H1
% in superimposed semilog and log-log plots

subplot(311)              % the first of 3 plots
semilogx(omega, abs(H1));
grid
title('Transfer function H1 in different log representations');
xlabel('Frequency / rad/s');
ylabel('Amplitude');

subplot(312)              % second of 3 plots
semilogy(omega, abs(H1));
grid
xlabel('Frequency / rad/s');
ylabel('Amplitude');

subplot(313)              % third of 3 plots
loglog(omega, abs(H1));
grid
xlabel('Frequency / rad/s');
ylabel('Amplitude');
```

The most appropriate representation for this example is the `loglog`
plot. In that case, the transfer function can be approximated with straight
line segments, which makes interpretation easier.

Do not be surprised that the labels partially overlap the graphs. That,
unfortunately, happens when multiple graphs are plotted on top of one
another using the `subplot` command. So here it often helps to enlarge
the graph to the full screen size.

We conclude this problem with the mention of a couple of very important errors. The following instructions in the definition of **H1** are very misleading:

```
H1 = 1/(j*omega);              % leads to a MATLAB error

H1 = 1/j*omega;                % syntactically correct, but still
                               % leads to a content error
```

In the first case we get an error message from MATLAB saying that here the matrix **1** must be divided by the matrix **j*omega**. What we should have calculated is the value of $\frac{1}{j\omega}$ for every frequency, (i.e., with respect to **omega** *term-by-term*). This is again a field operation.

The second error is somewhat more subtle, since MATLAB produced no error message. With this instruction, **1 / j** is multiplied by the vector **omega**. This is indeed syntactically correct, but it is not what should be calculated.

The third common error applies to the plot itself. The instruction

```
plot(omega,H1);
```

only generates a MATLAB warning, not an error. **H1** is a vector of *complex numbers* and can, therefore, not be plotted. MATLAB just plots the real part and gives a corresponding warning, which should be taken seriously in this case.

Solution to Problem 27 (file: none)
Regarding the labels, see the preceding solution to Problem 26. The rest of the problem is left to your experimental playfulness.

Solution to Problem 28 (file: solsurf.m)
The following instructions represent the important commands in the script file **solsurf.m**:

```
% Set up the vectors for the axis partitions

x = (-1:1:1);              % Grid in x direction
y = (-2:2:2)';             % Grid in y direction


% Constructing the grid matrices

v = ones(length(x),1);     % Auxiliary vector
X = v*x;                   % Grid matrix of x values
Y = y*v';                  % Grid matrix of y values


% Calculating the function values on the grid
```

```
f = sin(X.^2+Y.^2).*exp(-0.2*(X.^2+Y.^2));
```

For better understanding, it is interesting to have a closer look at the matrices X and Y:

```
>> X


X =
      -1     0     1
      -1     0     1
      -1     0     1

>> Y


Y =

      -2    -2    -2
       0     0     0
       2     2     2
```

Calculating $f(x, y)$ with the instruction

```
f = sin(X.^2+Y.^2).*exp(-0.2*(X.^2+Y.^2));
```

now relies on the fact that for each calculation the terms of X and Y will be combined *pairwise*. For example, the calculation involves $x = X_{11} = -1$ and $y = Y_{11} = -2$ for the term $(1, 1)$ of the function value matrix f. The matrix f thus contains the function value at the point $(-1, -2)$ there. All combinations for the lattice in the rectangle $[-1, 1] \times [-2, 2]$ with a grid separation of 1 or 2 will be run through in constructing the matrices X and Y.

Solution to Problem 29 (file: solsurf2.m)
The corresponding commands are named plot3, surf, mesh, and fill3. Put these commands[1] in the script file solsurf.m instead of surf and execute the script file by entering the command solsurf in the MATLAB command window.

Solution to Problem 30 (file: solTransFunc.m)
We limit ourselves to displaying the plot instructions from **solTransFunc.m**:

```
subplot(211)
loglog(omega, abs(H2));
grid
```

---

[1]We recommend consulting MATLAB help beforehand, in order to make correct syntactic use of them.

```
xlabel('Frequency / rad/s');
ylabel('Amplitude');

subplot(212)
semilogx(omega, angle(H2));
grid
xlabel('Frequency / rad/s');
ylabel('Phase');
```

Solution to Problem 31 (file: solexpfctplot.m)

An *overlay* plot on a graph is obtained with the commands

```
t = (0:0.01:2);
f1 = exp(-t/2);
f2 = exp(-2*t/5);
plot(t,f1,'b',t,f2,'r');
....                        % Labels
```

Graphs are placed next to one another or above and below one another using the `subplot` command, as with

```
subplot(121);       % Plot the first function
plot(t,f1,'b');
subplot(122);       % Plot the second function next to it
plot(t,f2,'r');
```

The plots can be copied into the cache using the menu command `Edit - Copy Figure`. Then they can be pasted into another windows application.

Solution to Problem 32 (file: sol2Dplot.m)

This problem can be solved as follows using the MATLAB function `meshgrid`:

```
% Setting up the vectors for partitioning the axes

x = (-2:0.2:2);            % Grid in the x-direction
y = (-1:0.1:1);            % Grid in the y-direction

% Constructing the grid matrices with meshgrid

[X,Y] = meshgrid(x,y);

% Calculating the function values on the grid

f = X.^2+Y.^2;
```

```
% Plot using the function surf

...
```

Solution to Problem 33 (file: none)

You can use the file **TestMat.txt**, which has been stored as a 3×3 matrix in ASCII format for solving this problem. Make sure that this file is stored in a search path of MATLAB.[2]

Next, enter the following sequence of commands in the MATLAB command window:

```
>> clear
>> load TestMat.txt
>> whos
>> TestMat
```

If everything has run properly, you should get the following responses:

```
>> whos
  Name           Size           Bytes  Class

  TestMat        3x3               72  double array

Grand total is 9 elements using 72 bytes

>> TestMat

TestMat =

     3    -4     5
    -1     4     6
     0     6    -3
```

The matrix of the text file can then be processed further in MATLAB.

With this technique you can conveniently upload data that is stored, for instance, in ASCII format into MATLAB for further processing.

Solution to Problem 34 (file: none)

You will find that it, unfortunately, *does not* work. Complex quantities must necessarily be processed in the form of real and imaginary parts that are read

---

[2]If in doubt, just enter the command **path**.

independently of one another. These can then be combined into a complex vector in the form

```
complexvector = realpartvector + j*imaginarypartvector;
```

Solution to Problem 35 (file: none)

With the instructions

```
>> clear
>> KVector = [1+j; i; 1-j]

KVector =

    1.0000 + 1.0000i
         0 + 1.0000i
    1.0000 - 1.0000i

>> save KVector
```

a complex vector is first defined and saved as a °.mat file in the Current Directory.

The sequence of commands

```
>> clear
>> load KVector
>> whos
  Name          Size              Bytes  Class

  KVector       3x1                  48  double array (complex)

Grand total is 3 elements using 48 bytes

>> KVector

KVector =

    1.0000 + 1.0000i
         0 + 1.0000i
    1.0000 - 1.0000i
```

shows that in this case the complex vector has been correctly stored and again read.

Unlike in the ASCII format (see Problem 34), complex fields can thus be stored and read in *binary* MATLAB format without difficulty.

Solution to Problem 36 (file: soldlmwrite.m)

```
% Define the matrix M

M = [   1   2   3; ...
      1+j   6   0; ...
      3+j   0  -j];

% Store the matrix with dlmwrite

dlmwrite('Mmatrix.txt', M,  'delimiter','\n','precision','%.4f');
```

Solution to Problem 37 (file: solaudio.m)

With the aid of MATLAB help, say by entering **help iofun**, you will find that **wavread** and **wavwrite** are the appropriate MATLAB commands for processing audio files in the °.wav format.

The following commands from **solaudio.m** load the file **Tada.wav**, display it graphically, multiply it by a factor of 10, and store the result again as a °.wav file:

```
% Load the wav-file Tada

[signal, samplerate, bits] = wavread('C:\windows\media\tada.wav');

% plot the signal

dt = 1/samplerate;          % calculate the sample interval
N = length(signal);         % find the number of points
time = (0:dt:(N-1)*dt);     % all time points from 0 to
                            % the time end point for the recording
                            %(this is (N-1)*dt)

plot(time, signal);         % Stereo signal contains two columns

% amplify the audio signal by a factor of 10

signal = 10*signal;

% store the new audio signal again as a *.wav file
% Note: the signal will be cut off if it overshoots
% by +1 or -1.  MATLAB provides appropriate warning messages.

wavwrite(signal, samplerate, bits,'C:\windows\media\tada10.wav');
```

Compare the two audio signals by playing them with a suitable program (e.g., with Windows Media Player).

Solution to Problem 38 (file: none)

```
;-)
```

Solution to Problem 39 (file: none)

First, the path `C:\mymatlab` is set up in the toolbar of the MATLAB command interface (see Fig. 1.1). This is done by pressing the ...-button and then selecting within the file choice window.

With that, the command

```
>> path(path, 'C:\mymatlab');
```

is added to the MATLAB search path (for commands and instructions) in the directory `C:\mymatlab`.

If the search path is to be kept for later MATLAB sessions, this must be done via the menu command `File - SetPath ...`. As an alternative to the above method, the new search path can be set here. After that, you can supply the search path with a directory containing all the subdirectories.

Solution to Problem 40 (file: none)

The complex terms in the (row) vector $\vec{r}$ are defined using the j symbol. For defining the corresponding column vector it is best to use the transposition operator `'`. Thus,

```
>> rz = [j j+1 j-7 j+1 -3];
>> rs = rz.';
```

Note that in the second row a *field operation* must be used.

Solution to Problem 41 (file: none)

```
>> V = repmat(M,2,2);
```

Solution to Problem 42 (file: sollogspace.m)

A look at MATLAB help (e.g., with `help elmat`) shows that the appropriate function for this is `logspace`. The sequence of commands is then

```
% create a logarithmically equidistant vector with logspace

v = logspace(0,1,10);

loglog(v,v);
```

The resulting graph is a straight line.

Solution to Problem 43 (file: solmeshgrid.m)

```
% set up the vectors for partitioning the axes

x = (-3:0.1:3);              % grid in the x-direction
y = (-3:0.1:3)';             % grid in the y-direction

% constructing the grid matrices with meshgrid

[X,Y] = meshgrid(x,y);

% calculating the function values on the grid

f = sin(X.^2+Y.^2).*exp(-0.2*(X.^2+Y.^2));
```

Solution to Problem 44 (file: none)

Naturally, the vector can be defined in MATLAB in the way described in Section 1.2.1 with

```
>> y = (1:0.1:10);
```

Alternatively, the command `linspace` can be used:

```
>> y = linspace(1,10,91);
```

This command is definitely better suited for calculating a certain *number* of *equidistant* reference points within a specified interval. Thus, in the above example you have, for instance, to consider that 91 reference points are needed in order to cover the interval [0, 10] exactly with a step size of 0.1; but this is inconvenient. In this case, that is, with a specified distance between the reference points, the first method is preferable.

If, on the other hand, you want an equidistant partition with a prespecified number of reference points, it is better to use `linspace`.

Solution to Problem 45 (file: none)

Calling `help elmat` shows that `fliplr` is a suitable function for this; hence,

```
>> y = (1:0.1:10);
>> yflip = fliplr(y)
yflip =

  Columns 1 through 11

   10.0000    9.9000    9.8000    9.7000    9.6000   ...
```

```
...

Columns 89 through 91

    1.2000    1.1000    1.0000
```

Solution to Problem 46 (file: solrepmat.m)

The instructions

```
% Define the vector z

z = (1:0.5:100);
factor = floor(length(z)/3);
rest = length(z)-3*factor;

% construct logical vectors each of which has a
% 1 in the third place

x = [0, 0, 1];                    % Auxiliary vector
Expandx = [repmat(x,1,factor), zeros(1,rest)];
xLogic = logical(Expandx);        % logical vector

% Extract data with the logical vector
% (and save as a column vector)

resultV = z(xLogic)'
```

first create an auxiliary vector with three components which are appended to one another **factor** times using **repmat**. This factor is calculated from the length of **z**. In order to make the factor integral, use the function **floor**.[3] The resulting numerical vector is then filled to the length of **z** with zeroes and converted into a logical vector. It will select the desired values.

This is shown by calling **solrepmat**:

```
>> solrepmat

resultV =

    2.0000
    3.5000
    5.0000
```

---

[3]In this example, the solution is generally well behaved. Of course, it was necessary to be able to figure out explicitly the required number of components and to enter it.

```
...
   98.0000
   99.5000
```

Solution to Problem 47 (file: none)

```
>> Graphic.yVals = [-2 0]

Graphic =

    title: 'Example'
   xlabel: 'time / s'
   ylabel: 'voltage / V'
      num: 2
    color: 'rb'
     grid: 1
    xVals: [0 5]
    yVals: [-2 0]
```

Other possibilities include

```
>> Graphic.yVals(1) = -2;
>> Graphic.yVals(2) = 0;
```

or

```
>> setfield(Graphic, 'yVals', [-2 0]);
```

Solution to Problem 48 (file: none)

```
>> color = Graphic.style.color(2)

color =

b
```

Solution to Problem 49 (file: solstructarray.m)

```
% definition of the structure "colors"

colors = struct('red', [], 'blue', [], 'green', []);

% defining a field of this structure using
% the function repmat

Colorfield = repmat(colors, 1, 20);
```

```
% the fields must be initialized
% in a loop

for k=1:20
    Colorfield(k).red = 'yes';
    Colorfield(k).blue = 'no';
    Colorfield(k).green = [0, 256, 0];
end
```

Calling solstructarray yields:

```
>> solstructarray
>> whos
  Name            Size                Bytes  Class

  Colorfield      1x20                 4472  struct array
  ans             1x25                   50  char array
  colors          1x1                   372  struct array
  k               1x1                     8  double array

Grand total is 249 elements using 4902 bytes


>> Colorfield

Colorfield =

1x20 struct array with fields:
    red
    blue
    green
```

Solution to Problem 50 (file: solelimInStructArray.m)

```
clear all

% defining the structure field

solstructarray

% eliminating the data field 'blue' in Colorfield

Colorfield2 = Colorfield;
clear Colorfield
for k=1:20
```

```
      Colorfield(k) = rmfield(Colorfield2(k), 'blue');
end

% change the value of 'green' in the 10th structure

Colorfield(10).green = [0, 256, 256];
```

Calling soleLimInStructArray yields

```
>> soleLimInStructArray
>> whos
  Name               Size                      Bytes  Class

  Colorfield         1x20                       3128  struct array
  Colorfield2        1x20                       4472  struct array
  colors             1x1                         372  struct array
  k                  1x1                           8  double array

Grand total is 384 elements using 7980 bytes

>> Colorfield

Colorfield =

1x20 struct array with fields:
    red
    green

>> Colorfield(10)

ans =

    red: 'yes'
  green: [0 256 256]
```

Solution to Problem 51 (file: none)

The attempt yields

```
>> clear
>> solstructarray
>> whos
  Name               Size                      Bytes  Class

  Colorfield         1x20                       4472  struct array
```

```
colors              1x1                         372  struct array
k                   1x1                           8  double array
```

```
Grand total is 224 elements using 4852 bytes
```

```
>> Colorfield(1) = rmfield(Colorfield(1), 'blue')
??? Subscripted assignment between dissimilar structures.
```

The reason is that the field elements have to be of the *same kind*. If the field
'**blue**' is eliminated in the first component, this structure does not match
the other 19. The elements of a field must, however, always be of the same
kind.

This is also the reason why the field **Colorfield** must first be cached
in the solution to Problem 50.

Solution to Problem 52 (file: solmeascampaign.m)

```
clear all

% (new) definition (reconstruction)
% of the cell array meascampaign
meascampaign{1,1} = {[256.9, 300.7], ...
                    [10, 27; 50, 16; 100, 5]};
meascampaign{1,2} = { [122.7, 103.1], ...
                    [5, 29; 50, 13; 100, 4; 200, 2]};
meascampaign{3,5} = { [101.0, 200.0], ...
                    [5, 31; 50, 22; 100, 12; 150, 6; 200, 1]};

% establish the dimension of the measurement value matrix

[n,m] = size(meascampaign{3,5}{2});

% set the measurement values

meascampaign{3,5}{2} = repmat([-1, 0],n,1);
```

The call

```
>> solmeascampaign
>> meascampaign{3,5}{2}

ans =

    -1    0
    -1    0
```

```
-1      0
-1      0
-1      0
```

shows that the measured values (in cell 2) of field element $(3, 5)$ have been
set in accordance with the problem statement.

Solution to Problem 53 (file: solweeklymeas.m)

```
clear all

% (new) definition (reconstruction)
% of the cell array measurements

measurements = {[256.9, 300.7], [10, 27; 50, 16; 100, 5]};

% extending the cell array by a
% third term
measurements(3) = {'H.E.Scherf'}

% Alternative: measurements{3} = 'H.E.Scherf'

% definition of the cell array weeklyMeasurements
% in which each cell contains the cell array measurements

for k=1:5
    weeklyMeasurements{k} = measurements;
end

% Check- output to the command window
weeklyMeasurements

% changing the name for each day

weeklyMeasurements{1}{3} = 'Engineer 1';
weeklyMeasurements{2}{3} = 'Engineer 2';
weeklyMeasurements{3}{3} = 'Engineer 3';
weeklyMeasurements{4}{3} = 'Engineer 4';
weeklyMeasurements{5}{3} = 'Engineer 5';

% Check- output to command window
weeklyMeasurements{1}
weeklyMeasurements{2}
weeklyMeasurements{3}
```

```
weeklyMeasurements{4}
weeklyMeasurements{5}
```

Calling

```
>> solweeklymeas

measurements =

    [1x2 double]    [3x2 double]    'H.E.Scherf'


weeklyMeasurements =

  Columns 1 through 4

    {1x3 cell}    {1x3 cell}    {1x3 cell}    {1x3 cell}

  Column 5

    {1x3 cell}



ans =

    [1x2 double]    [3x2 double]    'Engineer 1'


ans =

    [1x2 double]    [3x2 double]    'Engineer 2'


ans =

    [1x2 double]    [3x2 double]    'Engineer 3'


ans =

    [1x2 double]    [3x2 double]    'Engineer 4'


ans =

    [1x2 double]    [3x2 double]    'Engineer 5'
```

shows that `weeklyMeasurements` consists of a cell array in each component (cell) and that the name entries in these cell arrays (of type `measurements`) have been changed in each case via the last instructions.

Solution to Problem 54 (file: solweeklymeas2.m)

```
clear all

% (new) definition (reconstruction)
% of the cell array weeklyMeasurements using the
% script solweeklymeas

solweeklymeas;

% part (1) of the problem

weeklyMeasurements{5}{3} = 'A.E. Neumann'

% part (2) of the problem

weeklyMeasurements{1}{2} = [weeklyMeasurements{1}{2};[250,0]]

% part (3) of the problem

Day3 = weeklyMeasurements{3}

% part (4) of the problem

[Monday, Tuesday, Wednesday, ...
    Thursday, Friday]= deal(weeklyMeasurements{:})
```

Solution to Problem 55 (file: soltextscan.m)

```
% open the file with fopen

fid = fopen('sayings.txt');

% read in the first 7 words of the text with textscan

sayings = textscan(fid, '%s', 7);

% close the file with fclose
```

```
fclose(fid);

% display the result with celldisp
% and cellplot

celldisp(sayings)

% horizontal display forced with '
cellplot(sayings{:}')
```

Calling soltextscan yields

```
>> soltextscan

sayings{1}{1} =

All

sayings{1}{2} =

your

sayings{1}{3} =

base

sayings{1}{4} =

are

sayings{1}{5} =

belong

sayings{1}{6} =

to
```

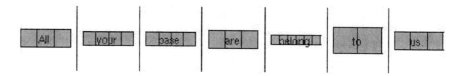

FIGURE 4.3   **Representing the cell array sayings with** `cellplot`.

```
sayings{1}{7} =

us.
```

and the graphic shown in Fig. 4.3.

Solution to Problem 56 (file: solfnExample3.m)

The vector of the squared components has to be included in the list of return vectors if it is to be accessed afterward in the MATLAB workspace. The function code for this looks like

```
function [sum, aquad] = solfnExample3(a, b)
%
% ....

sum = a+b;       % You have to do this.

aquad = a.^2;    % squaring of the components of a
```

It would have also been possible to include the vector $\vec{a}$ in the return value vector as well. But it is better programming style not to overwrite $\vec{a}$ with the function, since in later MATLAB commands, after execution of this function, this vector can be recalled if necessary.

Solution to Problem 57 (file: solfnExample2.m)

For brevity, here we only reproduce the major changes relative to `solfnExample2.m`:

```
function [t, sinfct, cosfct, expfct] = ...
                        solfnExample2(f1, Fa1, f2, Fa2, damp, Fa3)
%
% ...
%
%
% Input parameters:    ...
%
%                      Fa1, Fa2, Fa3
```

```
%                            Colors for sine, cosine, and exp function
%                            according to the plot syntax
% ...
%


... plot(t, sinfct, Fa1, t, cosfct, Fa2, t, expfct, Fa3) ...
```

Note that the color codes for **plot** must be passed on for a correct call of the function; for example,

```
>> [t, sinfct, cosfct, expfct] = ...
                      solfnExample2(3, 'r', 2, 'b', 1.5, 'g');
```

Solution to Problem 58 (file: solstructparam.m)

Here, again we only reproduce the major changes relative to **funcex2.m**:

```
function [t, sinfct, cosfct, expfct]= solstructparam(fctparams)
%
% Function solstructparam
%
% Call:  [t, sinfct, cosfct, expfct]= solstructparam(fctparams)
%
% ...


t=(0:0.01:2);
sinfct=sin(2*pi*fctparams.f1*t);
cosfct=2*cos(2*pi*fctparams.f2*t);
expfct=exp(-fctparams.damp*t);
...
```

The function call proceeds first by defining a corresponding parameter structure with the fields **f1**, **f2**, and **damp**. These must then be initialized with the desired values. These structures are then passed on as (single) parameters to the function **solstructparam**:

```
>> theParams.f1 = 4;
>> theParams.f2 = 2;
>> theParams.damp = -0.2;
>> [t, sinfct, cosfct, expfct]= solstrucparams(theParams);
```

Solution to Problem 59 (file: solplotcircle.m)

```
function [circumference, area]= solplotcircle(radius, color)
%
```

```
% ...
circumference = 2*pi*radius;
area = pi*radius^2;

% define the points on the circumference of the circle
% (with center 0)

alpha=(0:0.01:2*pi);     % angular partition in
                         % steps of 0.01 rad
x = radius*cos(alpha);
y = radius*sin(alpha);   % points on the circle in (x,y) coords

% plot polygonal sequence using fill in the specified colors
% and adjust axes (see the hint)

fill(x, y, color);
axis equal
```

The function can then be called, for example, with

```
>> [u, fl] = solplotcircle(3, 'r')
```

Solution to Problem 60 (file: solselpositiv.m)

```
function [outvect] = solselpositiv(vector)
%
% ...
% determine length N of input vector
% (for the loop end criterion)

N = length(vector);

% initialize output vector to the zero vector

outvect = [];

% select the positive elements with the loop

for k = 1:N              % run vector index from 1 to N
   if vector(k) > 0      % then attach an element to outvect
      outvect = [outvect,vector(k)];
   end;
end;
```

Solution to Problem 61 (file: solscalprod.m)

```
function [scprod] = solscalprod(vector1, vector2)
%
% ...

% determine length N of the input vectors
% (no check for equal lengths)

N = length(vector1);

% preinitialize the value of the scalar product to 0

scprod = 0;

% calculate the scalar product with loop

for k = 1:N                % run vector index from 1 to N
                           % form the product and add
                           % step-by-step to scprod
        scprod = scprod + vector1(k)*vector2(k);
end;
```

An experienced MATLAB programmer would certainly replace the loop with the command

```
vector1*vector2';
```

and thereby use the matrix algebra operations of MATLAB. Of course, the premise for this command is that the vectors are *row vectors* (with equal lengths).

Solution to Problem 62 (file: solfinput.m)

Essentially, the solution follows from the function **FInput** with different preinitialization of the output vector and a slight shift within the **while** loop.

```
function [InVector] = solfinput()
%
% ...

% preinitializing the output vector
% InVector to an EMPTY vector

InVector = [];
```

```
% read in the first data point

dat = input('Input a number.  (End if negative):');

% check and another input request until
% end criterion satisfied

while dat >= 0        % as long as the data point is positive;
                     % append to InVector
   InVector= [InVector; dat];
                     % read in the next number
   dat = input('Input a number.  (End if negative):');
 end;
```

Solution to Problem 63 (file: solcolorsin.m)

```
function [ ] = solcolorsin(color)
%
% ...

switch color
    case 'red'
       fcode = 'r'; % fcode is a variable for the
    case 'blue'      % plot color code
       fcode = 'b';
    case 'green'
       fcode = 'g';
    case 'magenta'
       fcode = 'm';
    otherwise
      disp('The input parameter was not allowed!');
      error('Please input help solcolorsin.');
  end

% plot function
t = (0:0.01:2*pi);
plot(t, sin(t), fcode);
```

Solution to Problem 64 (file: solroot2.m)

First, we consider the source code for the desired program:

```
function [approximation] = solroot2(startingvalue)
%
% ...
```

```
%

epsilon = 1e-3;         % 3 decimal places should be exact
xn = startvalue;        % initialize x_n to the starting value
xn1 = (startvalue^2+2)/(2*startvalue);
                        % approximating (x_n+1) with first
                        % iteration preinitializing

                        % testing whether the precision
                        % has been attained
while abs(xn1-xn)>epsilon
    oldx = xn1;         % for the next iteration save the
                        % old x_n+1 iteration formula,
                        % calculate new x_n+1
    xn1 = (xn1^2+2)/(2*xn1);
    xn = oldx;          % save old x_N+1 to new x_n
end;

approximation = xn1;
```

The desired precision is set by the preinstalled parameter **epsilon** 10^{-3}.

In each iteration step the difference between the new approximation to $\sqrt{2}$, namely **xn1**, is compared to the old approximation **xn** within the loop condition of the **while** loop. As long as the deviation exceeds 10^{-3}, iterations will continue.

Within the loop construct the value of the last approximation is first saved and then the next approximation is calculated.

It should be kept in mind that before entering into a **while**-loop, an approximation must initially be calculated with which the loop criterion can be tested.

Solution to Problem 65 (file: solwhile2.m)

```
function [res, rem] = solwhile2(a,b)
%
% Function solwhile2
%
% Call:    [res, rem] = solwhile2(a,b)
%
% example of call:  [res, rem] = solwhile2(10,3)
%
% ...
```

```
res = 0;
rem = a;

% computational loop

while rem >= b
   rem = rem - b;
   res = res + 1;
end;
```

Solution to Problem 66 (file: solstructin.m)

```
function [] = solstructin(f1,f2,grStruct)
%
% Function solstructin
%
% Call:    [] = solstructin(f1,f2,grStruct)
%
% Example of call: solstructin(1,0.5,grStruct)
%
% ...
% partition the plot range

t = linspace(grStruct.xVals(1), grStruct.xVals(2));

% plot graph

plot(t, sin(2*pi*f1*t), grStruct.color(1), ...
        t, sin(2*pi*f2*t), grStruct.color(2));

% label graph

title(grStruct.Title);
xlabel(grStruct.xlabel);
ylabel(grStruct.ylabel);

% set grid, if grid parameter 1

if grStruct.grid
    grid
end;

% set axis ranges
```

```
        axis([grStruct.xVals(1), grStruct.xVals(2), ...
            grStruct.yVals(1), grStruct.yVals(2)]);
```

Solution to Problem 67 (file: solvarargIn.m)

```
function [] = solvarargIn(time, varargin)
%
% Call:  solvarargIn(time,f1,f2,f3,...)
%
% Example of call: solvarargIn((0:0.1:2*pi),0.1,0.2,0.3)
%                  solvarargIn((0:0.1:2*pi),0.1,0.2,0.3,0.4)
%
% ...

% arrange frequencies in an array; varargin{:}
% can be set up as a list separated by commas.
% The [] then makes a numerical vector out of it.

freqs = [varargin{:}];
N = length(freqs);                    % set the number

% calculate the sine function and construct
% the plot argument vector (a matrix whose
% ROWS contain the sine values).

sfct = [];

for k=1:N
    sfct = [sfct; sin(2*pi*freqs(k)*time)];
end

% plot functions (color ordering occurs during plot)

plot(time, sfct)
```

Solution to Problem 68 (file: soleval.m)

The difficulty with this problem is that the parameter n is variable. Thus, neither the variable names nor the file names can be assigned *beforehand* in the program core. They have now to be *created dynamically as a string* using the current numbers. Here, the function **num2str** will be used to convert the number into a string.

This assignment can be forced within the program by means of the function **eval**:

```
function [] = soleval(n)
%
% Function soleval
%
% ...

for k = 1:n
    fname = ['Sig', num2str(k)];     % create signal names
                                     % create a string of commands
                                     % for the assignment
    assign = [fname, ' = ', 'rand(10,1);'];
    eval(assign);                    % execute command

    ffile = [fname, '.txt'];         % create file name
                                     % assemble the memory
                                     % command
    save_it = ['save ', ffile, ' ', fname, ' -ASCII'];
    eval(save_it);                   % execute save
end;
```

After execution of

```
>> soleval(5)
```

files **sig1.txt** to **sig5.txt** containing random vectors of length 10 should be found in the working directory.

In the above source text the following line should be specially noted:

```
save_it = ['save ', ffile, ' ', fname, ' -ASCII'];
```

ffile and **fname** are already strings and should not be between apostrophes. Spaces should also be provided, so the individual elements of the command string won't be attached to one other and appear as unknown commands to the MATLAB interpreter.

Solution to Problem 69 (file: fprinteval.m)

It is sufficient just to construct the *format string* inside the **for**-loop of the program **fprinteval** and then call the function **fprintf**:

```
...

for k=1:N
                                     % construct format string
```

```
    fstring = ['%', digs,'.', psts, 'f\n'];
    fprintf(fstring, x(k));
end
```

Solution to Problem 70 (file: soltrapez.m)

```
function [integral] = soltrapez(a, b, F, N)
%
% ...

h = (b-a)/(N);                  % subinterval length
intval = (a:h:a+N*h);           % reference points

                                % F at the interval boundaries
integral = (h/2)*(F(a)+F(b));

                                % evaluate F
                                % at the reference points
for i=2:1:N
    integral = integral+h*F(intval(i));
end;
```

Solution to Problem 71 (file: solquad.m)

Note that in the following solution the *function handle* **F** is used as a conventional function name in the command **fx = F(x)**:

```
function [] = solquad(a, b, F, N)
%
% ...

% call your own function

[integralT1] = soltrapez(a, b, F, N);  % trapezoidal rule
[integralS1] = fnExample4(a, b, F, N); % Simpson's rule

% call MATLAB functions

h = (b-a)/(N);                  % subinterval length
x = (a:h:a+N*h);                % reference points
fx = F(x);                      % function value

[integralTM] = trapz(x,fx);     % trapezoidal rule

[integralSM] = quad(F,a,b);     % Simpson's rule
```

```
% screen output

fprintf('\n');
fprintf('\nTrapez    (ours):  %12.8f', integralT1);
fprintf('\nTrapez  (MATLAB):  %12.8f', integralTM);
fprintf('\nSimpson   (ours):  %12.8f', integralS1);
fprintf('\nSimpson (MATLAB):  %12.8f\n\n', integralSM);
```

The solutions calculated using the trapezoidal rule are always consistent since they are based on the same reference points. A calculation with identical reference points was not possible for the Simpson rule, since the function **quad** has a different calling convention from **trapz**. But the solutions differ by little, as the following sample call shows:

```
>> solquad(0, 1, @sin, 4)
```

```
Trapez    (ours):    0.45730094
Trapez  (MATLAB):    0.45730094
Simpson   (ours):    0.45969832
Simpson (MATLAB):    0.45969769
```

Solution to Problem 72 (file: sollinpendde.m)

The differential equation must first be again converted into two first order differential equations. Using the procedure in Eq. (1.6) yields equations analogous to Eq. (1.8),

$$rl\dot{\alpha}_1(t) = \alpha_2(t) \, ,$$
$$\dot{\alpha}_2(t) = -\frac{g}{l} \cdot \alpha_1(t)$$

with the initial conditions

$$\vec{\alpha}(0) = \begin{pmatrix} \alpha(0) \\ \dot{\alpha}(0) \end{pmatrix} = \begin{pmatrix} \alpha_1(0) \\ \alpha_2(0) \end{pmatrix} \, . \tag{4.1}$$

This system of differential equations must be transformed as follows in MATLAB:

```
function [alphadot] = sollinependul(t, alpha)
%
% ...

    ...
```

```
                                    % first equation of first order
alphadot(1) = alpha(2);

                                    % second equation of first order
alphadot(2) = -(g/l)*alpha(1);
```

For initial values of $\frac{19\pi}{20}$ (a large deviation, nearly 180° relative to the equilibrium position) and 0 (initial velocity) the call reads as

```
>> [t, solutionLinear] = ode23(@sollinpendde, ...
    [0, 20], [19*pi/20,0]);
```

For a comparison, we can use the function **pendde** to calculate the solution of the nonlinear equation with the same initial conditions:

```
>> [t2, solution] = ode23(@pendde, [0, 20], [19*pi/20,0]);
```

Note that because of the inbuilt step size control in **ode23**, another time vector name must be used here. The lengths of **t** and **t2** are not necessarily the same.

Fig. 4.4, which was created using the commands

```
>> plot(t,solutionLinear(:,1),'r-',t2, solution(:,1),'b-')
```

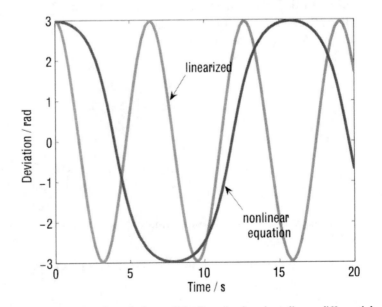

FIGURE 4.4 Comparing the solutions of the linearized and nonlinear differential equations for a mathematical pendulum with a large initial deviation.

```
>> xlabel('time / s')
>> ylabel('Deviation / rad')
```

reveals a distinct difference between the two solutions and, thereby, the error engendered by linearizing the differential equation for large deviations. With appropriate MATLAB calls, verify the good agreement between the two solutions for small deviations, say, $\frac{\pi}{8}$.

Solution to Problem 73 (file: solRCLP.m)

Excitation by a sinusoidal oscillation takes place directly on changing the file RCcomb.m, which can be done using a MATLAB function in this case:

```
function [udot]= solRCLP(t,u)
%
% ...
%

R = 10000;                  % resistance R
C = 4.7*10e-6;              % capacitance C of the capacitor
f = 3;                      % frequency of the driving frequency
                            % in Hz
                            % preinitialization
udot = 0;
                            % the first order equation
                            % u1(t) must be defined
                            % as a function in MATLAB

udot = -(1/(R*C))*u + (1/(R*C))*sin(2*pi*f*t);
```

Fig. 4.5 shows the result of the calculations for $f = 1$ Hz and $f = 3$ Hz produced by the MATLAB instructions

```
>> [t,solution] = ode23(@solRCLP, [0, 5], 0);
% now change the frequency from 1 to 3 Hz in the file solRCLP
>> [t2,solution] = ode23(@solRCLP, [0, 5], 0);
>> plot(t,solution(:,1),'r-',t2, solution2(:,1),'b-')
>> xlabel('time / s')
>> ylabel('Amplitude / V')
```

Transient effects, as well as the different damping of the amplitude and the phase shift, can be seen clearly.

The reader is encouraged to calculate the response of a low pass filter to a periodic rectangular pulse signal using the function rectangfunc in the accompanying software.

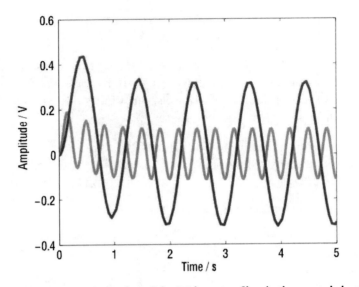

FIGURE 4.5 Response to excitation of the RC low pass filter in the example by two different oscillations.

Solution to Problem 74 (file: none)

For the case of a mathematical pendulum with a large initial deviation the sequence of instructions

```
>> [t1,solution1] = ode23(@pendde, [0, 20], [19*pi/20,0]);
>> [t2,solution2] = ode45(@pendde, [0, 20], [19*pi/20,0]);
>> plot(t1,solution1(:,1),'r-',t2, solution2(:,1),'b-')
```

reveals a difference between the solutions for values within the interval $[8, 20]$.

Solution to Problem 75 (file: solgrowth.m)

The differential equation is defined via the following function:

```
function [Pdot] = solgrowth(t,P)
%
% ...

alpha = 2.2;              % Parameter alpha
beta = 1.0015;           % Parameter beta

Pdot = [0];              % preinitializing

Pdot = alpha*(P^beta);   % first order equation
```

For numerical solution with MATLAB, a suitable initial condition must also be defined, say $P(0) = 1000$. The corresponding solution can then be calculated using

```
>> [t,P] = ode23(@solgrowth, [0, 30], 100);
```

Solution to Problem 76 (file: solPT1.m)

A careful look shows that here, to within the amplification factor K in Eq. (1.14), we are dealing with the equation of an RC low pass filter (1.11) with $T = RC$. Thus, taking $T = RC = 0.47$ and $K = 1$ and modeling the differential equation in the file solPT1.m with

```
vdot = (K/T)*u1(t)-(1/T)*v;
```

we find the unit step function response shown in Fig. 1.28 (see file u1.m).

Solution to Problem 77 (file: solparamODE.m)

The description of the solution procedures (solvers) in MATLAB help gives the following remarks about passing on additional parameters:

```
...

[T,Y] = solver(odefun,tspan,y0,options,p1,p2...) solves as
above, passing the additional parameters p1,p2... to the function
odefun, whenever it is called. Use options = [] as a place holder
if no options are set.

...
```

Thus, we first introduce a parameter for the length in the parameter list of the definition function (solparamODE) and, accordingly, pass on the code:

```
function [alphadot] = solparamODE(t, alpha, length)
%
% Function solparamODE
%
% ...

g=9.81;                        % acceleration of gravity m/s

                               % preinitializing
```

```
alphadot = [0;0];

                                % the first equation of first order
alphadot(1) = alpha(2);

                                % the second equation of first order
alphadot(2) = -(g/laenge)*alpha(1);
```

Then the differential equation can be solved as follows using **ode23**:

```
>> [t,solution] = ode23(@solparamODE, [0, 5], [pi/4,0], [], 20);
>> [t2,solution2] = ode23(@solparamODE, [0, 5], [pi/4,0], [], 5);
>> plot(t,solution(:,1),'r-',t2,solution2(:,1),'b-')
```

Here, the equation will obviously be solved twice with pendulum lengths of 20 and 5. The empty brackets ([ ]) are a placeholder for a structure of options (see **help odeset**), which must be specified at this point. In the present case the default options, which cover most cases, will be used because of this placeholder.

Fig. 4.6 shows the results of the calculations.

You can see clearly that the period is twice as long for the pendulum of length 20 as for the one with length 5 (period $= \frac{2\pi}{\sqrt{g/l}}$).

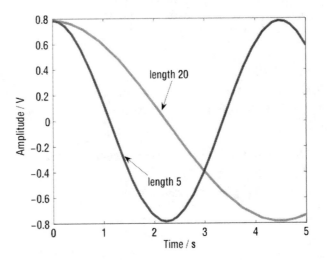

FIGURE 4.6 Solutions of the differential equation for the mathematical pendulum for pendulums of length 5 and 20.

Solution to Problem 78 (file: solDEsys.m)
The definition function for the system of differential equations has the
following form:

```
function [ydot] = solDEsys(t,y)
%
% Function solDEsys
%
% ...
                                    % preinitializing
ydot = [0;0];


                                    % first equation
ydot(1) = -2*y(1) - y(2);


                                    % second equation
ydot(2) = 4*y(1) - y(2);
```

The solution can then be calculated, say over the interval [0, 5], with

```
>> [t,solution] = ode23(@solDEsys, [0, 5], [1,1]);
>> plot(t,solution(:,1),'r-', t,solution(:,2),'b-')
```

This numerical solution can be compared with the exact solution using
the symbolics toolbox help and the function **dsolve**:

```
>> sol = dsolve('Dy1 = -2*y1 - y2', 'Dy2 = 4*y1 - y2', ...
                                'y1(0) = 1', 'y2(0) = 1', 'x')

sol =

    y1: [1x1 sym]
    y2: [1x1 sym]

>> pretty(sol.y1)

                       1/2              1/2
exp(- 3/2 x) (- 1/5 15    sin(1/2 15    x)
                                           1/2
                       + cos(1/2 15    x))
```

```
>> pretty(sol.y2)
```

$$- \frac{1}{2} \exp\left(- \frac{3}{2} x\right) \left(- \frac{6}{5} 15^{1/2} \sin\left(\frac{1}{2} 15^{1/2} x\right) - 2 \cos\left(\frac{1}{2} 15^{1/2} x\right)\right)$$

```
>> syms x
>> y1 = subs(sol.y1, x, t);
>> y2 = subs(sol.y2, x, t);
>> subplot(211)
>> plot(t,solution(:,1),'r-',t,solution(:,2),'b-')
>> xlabel('time / s');
>> subplot(212)
>> plot(t,y1,'r-',t,y2,'b-')
>> xlabel('time / s');
```

Note that the independent variable in the symbolic solution is called x in order to write over the default t. This is done so that the symbolic variable x can then be replaced by the time vector variable t used in the numerical solution. Although it appears in the output symbolic solution, the symbolic variable must be explicitly defined again before substitution with subs, since it is still unknown in the workspace. Fig. 4.7 shows that the numerical solution is obviously identical to the "exact" solution.

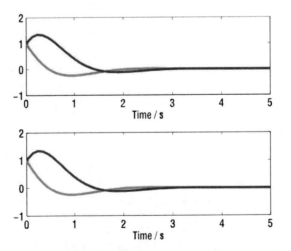

FIGURE 4.7   Solution of the system of differential equations from Problem 78. Top: numerical solution with ode23**, bottom: symbolic solution using** dsolve**.**

Solution to Problem 79 (file: soldeval.m)
Calling the following MATLAB code

```
% solution of the pendulum differential equation with ode23

solStruct = ode23(@pendde, [0, 20], [pi/4,0])

% evaluation using deval with a step size of 0.01
refpoints = (solStruct.x(1):0.01:solStruct.x(end));
sol = deval(solstruct, refpoints)

% plot the solution (caution:  row vectors!)

plot(refpoints, sol(1,:), 'r-', refpoints, sol(2,:), 'g--')
```

with **soldeval** yields

```
>> soldeval

solStruct =

     solver: 'ode23'
    extdata: [1x1 struct]
          x: [1x95 double]
          y: [2x95 double]
      stats: [1x1 struct]
      idata: [1x1 struct]


sol =

  Columns 1 through 5

      0.7854      0.7854      0.7853      0.7851      0.7848
           0     -0.0069     -0.0139     -0.0208     -0.0277

  Columns 6 through 10

      0.7845      0.7841      0.7837      0.7832      ...
     -0.0347     -0.0416     -0.0485     -0.0555      ...
```

We omit displaying the plots here.

Solution to Problem 80 (file: solsymint.m)

```
% define x as symbolic quantity

syms x

% define the function g(x)

g = sin(5*x-2);

% integrate the function g(x) once, symbolically

G = int(g,x);

% integrate the function G(x) symbolically

G2 = int(G,x);
```

Next, it is interesting to take a look at the workspace after these instructions. This gives

```
>> whos
  Name      Size           Bytes  Class

  G         1x1              154  sym object
  G2        1x1              156  sym object
  g         1x1              144  sym object
  x         1x1              126  sym object

Grand total is 46 elements using 580 bytes
```

You can see from the entry **sym object** that the entire set of quantities are symbolic quantities.

The calculated functions are

```
>> G

G =

-1/5*cos(5*x-2)

>> G2

G2 =

-1/25*sin(5*x-2)
```

Note that the *constants of integration have not been taken into account*.

Solution to Problem 81 (file: solsymtaylor.m)

```
% define x as a symbolic quantity

syms x

% define the function g(x)

g = sin(5*x-2);

% calculate Taylor expansion of the function g(x)

T3 = taylor(g,4,1);
pretty(T3)
```

yields

$$\sin(3) + 5 \cos(3) (x - 1) - 25/2 \sin(3) (x - 1)^2$$
$$- 125/6 \cos(3) (x - 1)^3$$

Note that the second parameter must be set at one higher than the degree of the Taylor polynomial which it is to determine.

Solution to Problem 82 (file: solsymDE.m)

```
% define x as a symbolic quantity

syms x

% define the differential equation (a STRING!)

theDE = 'Dy = x*y^2';

% calculate the solution using dsolve

sol = dsolve(theDE, 'x')
```

This yields the solution

```
sol =

-2/(x^2-2*C1)
```

Solution to Problem 83 (file: solysymDiff.m)

```
% define x and g(x) as symbolic

syms x
g = sin(5*x-2);
% take the derivative of the function g(x) twice, symbolically

g1 = diff(g,x);
g2 = diff(g,x,2);

% establish the time vector for the plot

t=(0:0.01:10);

% Make numerical vectors out of the symbolic
% functions, by replacing the symbolic variable
% x by the vector t
% (note: name the variables differently, in order
% to avoid overwriting)

G  = subs(g,x,t);
G1 = subs(g1,x,t);
G2 = subs(g2,x,t);

% plot the result (G in red, G1 in blue, G2 in green)

plot(t,G,'r-', t,G1,'b', t,G2,'g');
```

Solution to Problem 84 (file: solsymMaple.m)

```
% define x and g(x) as symbolic
% taking MAPLE syntax into account

syms x
maple('g := sin(5*x-2);');

% integrate the function g(x) twice, symbolically

maple('G:= int(g,x);');
G2 = maple('int(G,x);')
```

4.2 SOLUTIONS TO THE SIMULINK PROBLEMS

Solution to Problem 85 (file: s_test1.mdl)

When doing comparative calculations, it is important, first of all, to make sure all the parameters are set correctly. In particular, the simulation duration, step size, and number of results to be returned to the MATLAB workspace in the **Scope** block under the variables **S_test1_signals** must fit.

Thus, for example, for a fixed step size of 0.00001 and a simulation duration from 0 to 10, it must be kept in mind that a total (of three times) 1000001 values of **S_test1_signals** also have to be passed to MATLAB. This is guaranteed if the parameter **Limit data points to last** in the file window **Scope Properties - Data history** of the **Scope** block is set for at least 1000001 values. The MATLAB sinks of Simulink have a similar limitation on the output.

To set the match for the step size to 0.00001 in the case where a procedure with a step size control is chosen (e.g., **ode23**), the Output options entry in the **Configuration Parameters** file window must be reset to **Produce specified output only**. The desired time points then have to be specified specifically under **Output times**, in the present case to **[0:0.00001:10]**.

In sum, the three procedures include one that provides fixed settings of the step size to 0.00001 using **ode3**, one with step size control using **ode23** and an output with step size of 0.00001, and one with step size control using **ode23** alone. All three yield graphically identical results in the present example. In the third case, however, the simulation time is clearly shorter than in the first two cases.

Solution to Problem 86 (file: none)

Let us call the output of the integrator block $y(t)$, so that the input signal $x(t)$ and the output signal are related by

$$y(t) = \int\limits_0^t x(\tau)\,d\tau + y(0)\,.$$

Differentiation of this equation gives Eq. (2.1). The system therefore solves this (somewhat degenerate) differential equation.

Solution to Problem 87 (file: s_soldiff.mdl)
Like the integrator block, the differentiator block appears in the function library Continuous. Essentially, you obtain s_soldiff by simply replacing the integrator block with the derivative block in s_test1.

Solution to Problem 88 (file: s_solIVP.mdl)
At this point, this problem naturally anticipates Section 2.3.

In light of Problem 86 it can be argued that basically you only have to replace the input signal $x(t)$ by $-2u(t)$. Thus, on integrating $-2u(t)$ you get $u(t)$. The general theoretical solution raises the difficulty that $u(t)$ is not known, but this is actually the principle of the solution. You simply feed the output, with a factor of -2, back to the input of the integrator and leave the rest to Simulink or, rather, to the numerical solution procedure used by Simulink.

This leads to the system shown in Fig. 4.8. To see the signal under MATLAB, use the following plot command.

```
>> plot(solIVP(:,1), solIVP(:,2));
```

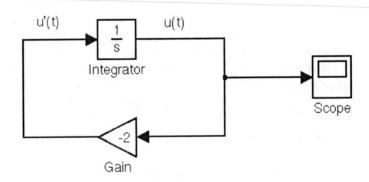

FIGURE 4.8   Simulink system for solving $\dot{u}(t) = -2 \cdot u(t)$.

The initial value is fixed in the integrator block parameter window by setting the parameter initial condition.

The reader should compare the simulated solution (e.g., in the interval $[0, 3]$) with the exact solution $u(t) = e^{-2t}$ using MATLAB.

Solution to Problem 89 (file: s_solOscill.mdl)
The modified medium and the drag coefficient are appropriate for use in Eq. (2.11) for the coefficient of friction b.

The density of water is known[4] to be 1000 kg/m^3.

Thus, for the case in which the flat side is in the direction of motion, we have

$$b_1 = 1.2 \cdot \frac{1}{2} \cdot 63.6943 \cdot 10^{-4} \cdot 1000 \ \mathrm{m}^2 \frac{\mathrm{kg}}{\mathrm{m}^3} = 3.8217 \ \frac{\mathrm{kg}}{\mathrm{m}}$$

and in the case in which the round side is in the direction of motion,

$$b_2 = 0.4 \cdot \frac{1}{2} \cdot 63.6943 \cdot 10^{-4} \cdot 1000 \ \mathrm{m}^2 \frac{\mathrm{kg}}{\mathrm{m}^3} = 1.2739 \ \frac{\mathrm{kg}}{\mathrm{m}} \ .$$

We assume that the round side is in the positive direction of motion[5] and the flat side, in the negative direction, so that b_1 is used if $sgn(x(t)) \geq 0$ and b_2 if $sgn(x(t)) < 0$.

This can be done using the **switch** block.

Fig. 4.9 shows the segment of the system **s_solOscill** of interest.

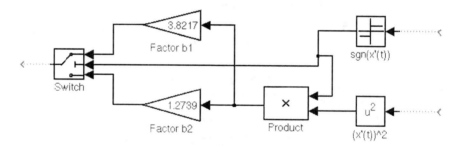

FIGURE 4.9   Detail from the Simulink system s_solOscill.

Here, the **switch** block is controlled by the output of the *sgn* function of the installed **Fcn** block from the **User-defined Functions** function library. The parameter **Threshold** controls the threshold according to which a high or low input is switched on.

In conclusion, it should be noted that the substitution corresponding to Fig. 4.9 is easier if the **Fcn** block is used to calculate the whole quantity $b \cdot sgn(x(t)) \cdot x^2(t)$ (see the solution to Problem 90). You just have to remember that the input signal for this block is *always* called **u**.

---

[4]A liter of water has a mass of 1 kg and 1 m^3 contains 1000 liters.
[5]This is a matter of definition. *You* have to establish the direction of motion in the model.

Solution to Problem 90 (file: s_solpendul.mdl)

Fig. 4.10 shows the solution of the differential equation for the mathematical pendulum implemented in the system **s_solpendul**.

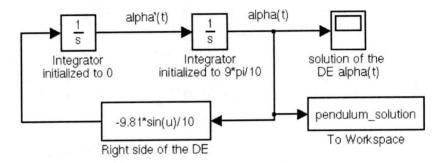

FIGURE 4.10   Simulink system **s_solpendul** for solving the (unlinearized) differential equation for the mathematical pendulum.

Solution to Problem 91 (file: s_solDE2ord.mdl)

Fig. 4.11 shows the Simulink system that solves the initial value problem from Eq. (2.13) with the perturbation function e^{-t}.

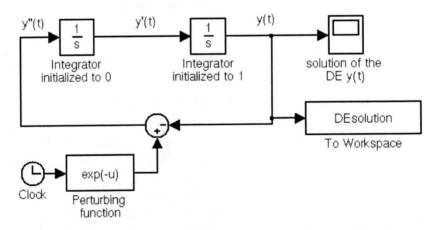

FIGURE 4.11   Simulink system for solving the initial value problem, Eq. (2.13).

The implementation of the perturbation function is of some interest here. A **Clock** block, which essentially provides the time argument **t**, drives a **Fcn** block, which has been set to **exp(-u)**. In this way the additive perturbation term e^{-t} is generated in the right-hand side of the differential equation.

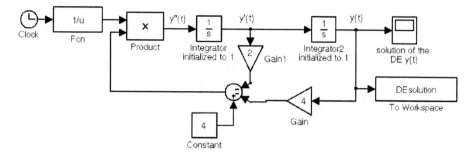

FIGURE 4.12   Simulink system for solving the initial value problem (2.14).

Solution to Problem 92 (file: s_solTD1.mdl)

First, Eq. (2.14) has to be reduced to the form

$$\ddot{y}(t) = \frac{1}{t}(4 - 2\dot{y}(t) - 4y(t)), \qquad y(1) = 1, \ \dot{y}(1) = 1 \qquad (4.2)$$

Fig. 4.12 shows the Simulink system, which solves the initial value problem. It is a graphical reflection of Eq. (4.2).

Note that the simulation start point must be set to 1 (viz. the initial values!).

Solution to Problem 93 (file: s_solTD2.mdl)

As opposed to the examples shown in Section 2.3, this case involves solving a *system* of (linear) differential equations.

This sort of system of equations can, in fact, be solved using Simulink just as in the case of a single differential equation. The sole difference is that the equations are coupled, a feature which is embodied in the corresponding connections between blocks in Simulink.

Fig. 4.13 shows a possible Simulink solution for the initial value problem of Eq. (2.15).

Here, you can see, for example, that the signal **y1'(t)** results from the addition of the signals **-2y2(t)** and **-3y1(t)**. This corresponds exactly to the first of Eq. (2.15).

This system of differential equations can be solved in closed form. The solution can be determined using the command **dsolve** in the following way:

```
>> syms y1 y2              % define y1 and y2
                           % as symbolic variables
```

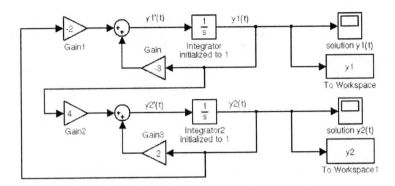

FIGURE 4.13   Simulink system for solving the system of differential equations (2.15).

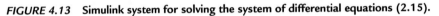

```
>> % calling the command dsolve for systems (see helpdsolve)
>>
>> S = dsolve('Dy1 = -3*y1 -2*y2','Dy2 = 4*y1 + 2*y2', ...
                             'y1(0) = 1','y2(0) = 1')


S =

    y1: [1x1 sym]
    y2: [1x1 sym]

>> Y1 = S.y1;
>> Y2 = S.y2;
>> pretty(simplify(Y1))

                              1/2        1/2           1/2
- 1/7 exp(- 1/2 t) (-7 cos(1/2 t 7    ) + 9 7    sin(1/2 t 7    ))

>> pretty(simplify(Y2))

                         1/2              1/2            1/2
1/7 exp(- 1/2 t) (13 7     sin(1/2 t 7    ) + 7 cos(1/2 t 7    ))
```

This yields two damped oscillators as solutions:

$$rly_1(t) = -\frac{1}{7}e^{-\frac{1}{2}t}\left(-7\cos\left(\frac{\sqrt{7}}{2}t\right) + 9\sqrt{7}\sin\left(\frac{\sqrt{7}}{2}t\right)\right),$$

$$y_2(t) = \frac{1}{7}e^{-\frac{1}{2}t}\left(7\cos\left(\frac{\sqrt{7}}{2}t\right) + 13\sqrt{7}\sin\left(\frac{\sqrt{7}}{2}t\right)\right).$$

If the system s_solTD2 is simulated with the set parameters (fixed step size 0.01, time interval [0, 10]), then the following vectors appear in the MATLAB workspace:

```
? whos
  Name        Size          Bytes  Class

    S          1x1            770   struct array
   time       901x1          7208   double array
   Y1          1x1            262   sym object
   Y2          1x1            260   sym object
   y1         901x1          7208   double array
   y2         901x1          7208   double array


Grand total is 2983 elements using 22916 bytes
```

In order to be able to compare the signals with one another, you substitute the time vector time created by the system s_solTD2 for t in the symbolic variables Y1 and Y2 as follows:

```
>> syms t                  % symbolic variables in Y1 and Y2
>> sol1 = subs(Y1,t,time);% replace t with time values
>> sol2 = subs(Y2,t,time);% replace t with time values
```

We can now compare the numerical Simulink solutions y1 and y2 with the exact solutions sol1 and sol2. We do this by comparing the columns of numbers with one another, since, because of the expected small differences, a comparison of the graphs would not be meaningful.

```
>> [sol1(1:25),y1(1:25)]   % displaying the first 25 values

ans =

     1.0000    1.0000
     0.9502    0.9502
     0.9006    0.9006
     0.8514    0.8514
     0.8025    0.8025
     0.7539    0.7539
     0.7056    0.7056

      ...
    -0.0162   -0.0162
    -0.0580   -0.0580
    -0.0993   -0.0993
```

In fact, no differences can be seen within the first 25 values. Here, we omit the corresponding check for the second solution.

As a test, the reader could also check the solutions using appropriate plot commands.

Solution to Problem 94 (file: s_DEnon4.mdl)
The solution to this problem is illustrated in Fig. 4.14.

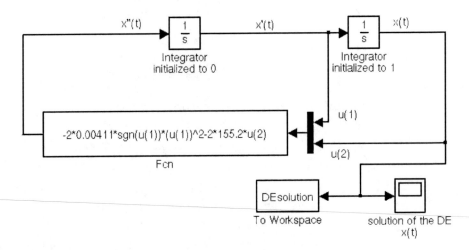

FIGURE 4.14 **Simulink system for solving Eq. (2.12).**

Here, the signals $x(t)$ and $\dot{x}(t)$ are first combined into a vector signal by a multiplexer. The differential equations is realized within the **Fcn** block whereby the individual components of the vectorial input signal are processed. Most of the blocks in the original system **s_Denon** are left out.

Solution to Problem 95 (file: s_solTD3.mdl)
The system of differential equations (2.15) can be solved using the **Fcn** blocks shown in Fig. 4.15.

Here, the right-hand sides of the equations are each completely realized in the **Fcn** blocks and the solutions are assembled into a vector valued signal, which is used as an input signal for the **Fcn** blocks.

Solution to Problem 96 (file: s_DEnon5.mdl)
The result of combining the blocks in the feedback branch is shown in Fig. 4.16.

It should first be noted that this subsystem has *two* inputs, unlike the example from Section 2.4. This construction may be rather difficult to carry

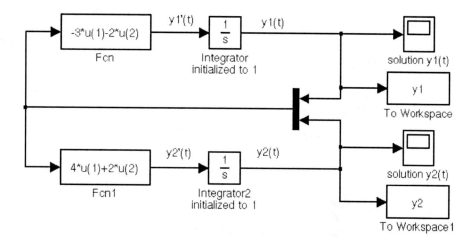

FIGURE 4.15 Simulink system for solving the system of differential equations (2.15).

out if the blocks are also connected to the integrators and sink blocks. As long as the system setup is incorrect, you will find that the menu entry `Edit - Create Subsystem` cannot be selected. When that happens, you have to recheck the selected partial system.

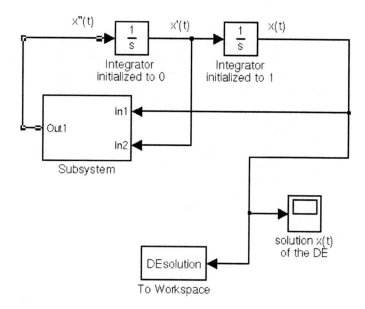

FIGURE 4.16 The system `s_denon5.mdl`.

Solution to Problem 97 (file: none)

In the Simulink system **s_pendul** the length of the pendulum and the acceleration of gravity are processed for the gain parameter in the **Gain** block in the form **-9.81/10**.

The block is named **Gain** and the parameter to be changed is likewise called **Gain**. This is shown by calls with **find_system** and **get_param**:

```
>> blks = find_system('s_pendul', 'Type', 'block')

blks =

    's_pendul/Gain'
    's_pendul/Integrator1'
    's_pendul/Integrator2'
    's_pendul/sin(u)'
    's_pendul/Out1'

>> GainPars = get_param('s_pendul/Gain','DialogParameters')

GainPars =

        Gain: [1x1 struct]
        Multiplication: [1x1 struct]
        ...
```

Note that the Simulink system **s_pendul** has to be open in order for **find_system** to work reasonably. Otherwise, an error message appears.

The corresponding calls for setting a new pendulum length as well as the calls for the system and plotting the results look like

```
>> set_param('s_pendul/Gain','Gain','-9.81/5');
>> [t,x,y] = sim ('s_pendul', [0, 30]);
>> set_param('s_pendul/Gain','Gain','-9.81/8');
>> [t1,x1,y1] = sim ('s_pendul', [0, 30]);
>> plot(t,y,'r-',t1,y1,'b:');
```

Here we omit a display of the graph.

Solution to Problem 98 (file: solsimpendul.m)

```
function [t,Y] = solsimpendul(lng)
%
% ...
%
% initializing the output matrix Y as empty
```

```
% and the iteration duration

Y = [ ];
iterations = length(lng);

% calling the iteration loop for the simulation
% In this version of the program you must make sure
% that a fixed step procedure is set
% in the parameter window

for i=1:iterations
    % Set the gain parameter with set_param
    % for each new iteration!
    frq = num2str(-9.81/lng(i));
    set_param('s_Pendul/Gain','Gain',frq);

    [t,x,y] = sim('s_Pendul', [0,30]);
    Y = [Y,y];
end;
```

The following is an example of a call with a plot of the results:

```
>> [t,Y] = solsimpendul([2,5,8,10]);
>> plot(t,Y)
```

Please note that for this version of the program a fixed step procedure must be set in **s_pendul**. Moreover, the Simulink system must be *open*, otherwise MATLAB responds with an error message in a form something like

```
?? Error using ==> set_param
Invalid Simulink object name: s_pendul/Gain.

Error in ==> solsimpendul.m
On line 43  ==>    set_param('s_pendul/Gain','Gain',frq);
```

Solution to Problem 99 (file: solDEnonit.m)
Here, we show only the changes relative to the source text of **Denonit**:

```
function [t,Y] = solDEnonit(r, rho, Fc, sz, step, anfs)
%
% ...
%
% initializing the output matrix Y as empty
% and the iteration length
```

```
Y = [ ];
iterations1 = length(r);
iterations2 = length(Fc);  % new relative to DEnonit

% calling the iteration loop ...

for k=1:iterations2  % new relative to DEnonit

        for i=1:iterations1
          ...
        c = num2str(Fc(k));      % new relative to DEnonit
        set_param('s_DEnon3/Factorc','Gain',c);
        [t,x,y] = sim('s_DEnon3', [0,sz]);
        Y = [Y,y];
    end;

end;  % new relative to DEnonit
```

An example of a call (with the Simulink system **s_DEnon3** open) with subsequent plotting of the results looks like

```
>> [t,Y] = solDEnonit([3.0 20.0], 1.29/1000, ...
                      [155.2 205.2], 5, 0.01, [0,1]);
>> plot(t,Y)
```

We again omit a display of the graphic here.

Solution to Problem 100 (files: s_solsin.mdl, solsimsin.m)

The Simulink system **s_solsin** consists just of a **sine wave** block and an **outport**, so that we can omit the graphical representation.

The system can be called via the following function for a frequency vector:

```
function [t,Y] = SolSimSin(frequencies)
%
% ...
%Y = [ ];
iterations = length(frequencies);

% ...

for i=1:iterations
   % set the gain parameters with set_param
   % for each new iteration!
   frq = num2str(frequencies(i));
   set_param('s_SolSin/Sinefunction','Frequency',frq);
```

```
    [t,x,y] = sim('s_SolSin');
    Y = [Y,y];
end;
```

An example of a call (with the Simulink system **s_solsin** open) with subsequent plotting of the results looks like

```
>> [t,Y] = SolSimSin(2*pi*[0.1, 0.3]);
>> plot(t,Y)
```

Solution to Problem 101 (files: s_solSimDE.mdl, solSimDE.m)
Fig. 4.17 shows the Simulink system for solving the initial value problem of Eq. (2.17).

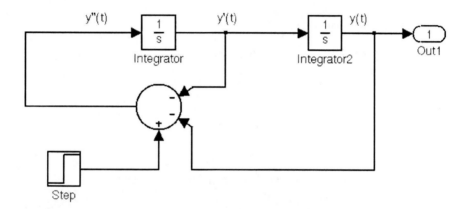

FIGURE 4.17   Simulink system for solving the initial value problem (2.17).

The system can be called via the following function for variable step heights h:

```
function [t,Y] = solSimDE(heights)
%
% ...

    ...

for i=1:iterations
    % Set the gain parameters with set_param
    % for each new iteration.
    h = num2str(heights(i));
    set_param('s_solSimDE/Step','After',h);
```

```
      [t,x,y] = sim('s_solSimDE');
      Y = [Y,y];
end;
```

An example of a call (with the Simulink system **s_solsimDE** open) with subsequent plotting of the results looks like

```
>> [t,Y] = SolSimDE([1,2,3]);
>> plot(t,Y)
```

Solution to Problem 102 (files: s_DEnon6.mdl, solDEnonit2.m)
Here, the trick is to use a *cell array* as the output parameter Y rather than a numerical field. Here, we reproduce the most important instructions relevant to this from the file **solDEnonit2.m**. All the other instructions are identical to those in **denonit.m**.

```
function [T,Y] = solDEnonit2(r, rho, Fc, sz, step, anfs)
%
% Function  solDEnonit2

% ...


...

% Initializing the output parameters T and Y
% as empty CELL ARRAYS.

T = {}; Y = {};

% Calling the iteration loop for the simulation
% (iteration duration by calling sim and
% the fixed step size from 0.0 to sz)

for i=1:iterations
   % Set the block parameters with set_param
   % for each new iteration.
   Reciprocalmass = num2str(1/M(i));
   set_param('s_DEnon6/invm','Gain',Reciprocalmass);
   b = num2str(B(i));
   set_param('s_DEnon6/Factorb','Gain',b);
   c = num2str(Fc);
   set_param('s_DEnon6/Factorc','Gain',c);
   [t,x,y] = sim('s_DEnon6', [0,sz]);
```

```
    % Extend the cell arrays by one (cell-) component.
    T = [T,{t}];
    Y = [Y,{y}];
end;
```

A call with the instructions

```
>> clear all
>> r=[3.0, 20.0, 60.0];          % Radius vector
>> rho=1.29*1000/1000000;        % rho stays the same each time
>> Fc=155.2;                      % c stays the same each time
>> [t,Y] = solDEnonit2(r, rho, Fc, 5, 0.01, [0,1]);
```

yields two 1×3 cell arrays in the workspace:

```
>> whos
  Name      Size                   Bytes  Class

  Fc        1x1                        8  double array
  Y         1x3                     1764  cell array
  r         1x3                       24  double array
  rho       1x1                        8  double array
  T         1x3                     1764  cell array

Grand total is 216 elements using 1884 bytes
```

An analysis of the cell entries shows that the numerical vectors in the cells have different lengths:

```
>> Y

Y =

    [161x1 double]    [27x1 double]    [10x1 double]

>> T

T =

    [161x1 double]    [27x1 double]    [10x1 double]
```

The result can be visualized graphically with

```
>> plot(T{1},Y{1},'b-', T{2},Y{2},'k--', T{3},Y{3},'r-.')
```

On calling the plot command, you will see that the curves are somewhat "jagged," since few reference points were used. The interpolation procedures described above should be used to get a smoother display.

We skip the graphical display of the results here.

Solution to Problem 103 (file: s_solcharc.mdl)

Fig. 4.18 illustrates the Simulink system to be designed.

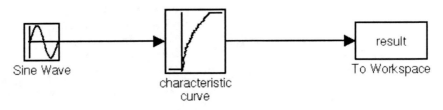

FIGURE 4.18   Simulink system for simulation of the characteristic curve (2.21).

In order to parametrize the system correctly, the characteristic curve first has to be defined via MATLAB in a form suitable for the characteristic curve block. We set x and $k(x)$ in two MATLAB vectors. Since the function $k(x)$ is defined piecewise, this procedure must also be done step-by-step:

```
>> x1 = (-2:0.1:0);      % negative x axis to -2
>> x2 = (0:0.1:4);       % range where k(s)= sqrt(x)
>> x3 = (4:0.1:6);       % range from 4 to 6
>> y1 = zeros(1,length(x1));
>> y2 = sqrt(x2);         % determine the function values
>> y3 = 2*ones(1,length(x3));
>> x = [x1,x2,x3];        % set up vectors
>> k = [y1,y2,y3];
```

Note that the characteristic curve block *extrapolates* the values outside the defined segments, in this case to 0 at the bottom and to 2 at the top, as the definition of $k(x)$ also specifies.

As you can see, the preceding commands have already been used in a problem in Chapter 1 on MATLAB techniques.

After these instructions have been entered into MATLAB, the characteristic curve can be seen in the characteristic curve block, if the vectors **x** and **k** have been entered appropriately there. In addition, we enter the variable **amp** in the amplitude parameters of the **sine wave** block in order to be able to change the amplitude from MATLAB.

The simulation for amplitudes 1, 3, and 5, with

```
>> amp=1;                          % Set amplitude to 1
>> sim('s_solcharc.mdl');          % Simulate the Simulink system
>> res1=res;                       % Save result
>> amp=3;                          % Set amplitude to 3
>> sim('s_solcharc.mdl');          % Simulate the Simulink system
>> res3=res;                       % Save result
>> amp=5;                          % Set amplitude to 5
>> sim('s_solcharc.mdl');          % Simulate the Simulink system
>> res5=res;                       % Save result
>> plot(t,res1,'r-', t,res3,'b:', t,res5,'g--');
```

yields the result shown in Fig. 4.19.

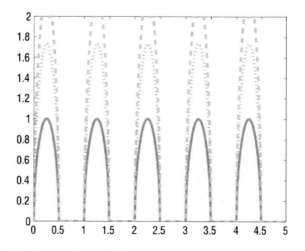

FIGURE 4.19 Simulation of s_solcharc with amp=1, 2, and 5.

You can see that negative values of the sine have been set equal to 0 with the characteristic curve. If the amplitude exceeds 4, the output value is set at 2 for the given time. In between, the value of $\sqrt{a}\sin{(2\pi x)}$ is plotted.

Solution to Problem 104 (file: s_solcharfam.mdl)

Fig. 4.20 shows the Simulink system to be designed.

For this system, the parameters of the family of characteristic curves, x, y, and k, set in s_solcharfam have to be set using the following MATLAB instructions:

```
>> x = (-1:0.1:1);
```

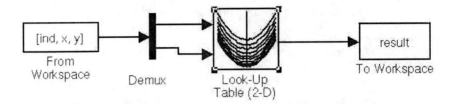

FIGURE 4.20   **Simulink system for simulating the family of characteristic curves (2.22).**

```
>> y = (-1:0.1:1);
>> ind = (0:1:length(x)-1);
>> [X,Y] = meshgrid(x,y);
>> k = X.^2+Y.^2;
```

The parameters of the `From Workspace` block, `ind`, `x`, and `y` must *under all circumstances* be *column vectors*. This is done using the following MATLAB commands:

```
>> x = x';
>> y = y';
>> ind = ind';
```

As expected, the simulation yields a graph of the function $2 \cdot x^2$ in the interval $[-1, 1]$. Here, we omit the display of this result (with `plot(x,res)`).

TABLE OF ARITHMETIC MATLAB OPERATIONS

The following tables summarize the application of arithmetic operations in MATLAB as matrix and field operations.

Here, we shall use the matrices

$$A = \begin{pmatrix} 2 & 1 \\ 1 & 1 \end{pmatrix}, \qquad B = \begin{pmatrix} -1 & 1 \\ 1 & 1 \end{pmatrix}$$

and the vectors

$$\vec{a} = \begin{pmatrix} 2 \\ 1 \end{pmatrix}, \qquad \vec{b} = \begin{pmatrix} -1 \\ 1 \end{pmatrix}$$

as examples of matrices and vectors.

A.1   Arithmetic Operations as Matrix Operations

The first table summarizes the arithmetic operations of MATLAB as *matrix operations*.

Operation in MATLAB	Effect	Example
A+B	matrix addition	$A + B = \begin{pmatrix} 1 & 2 \\ 2 & 2 \end{pmatrix}$
A−B	matrix subtraction	$A - B = \begin{pmatrix} 3 & 0 \\ 0 & 0 \end{pmatrix}$
a+b	vector addition	$\vec{a} + \vec{b} = \begin{pmatrix} 1 \\ 2 \end{pmatrix}$
A*B	matrix multiplication	$A \cdot B = \begin{pmatrix} -1 & 3 \\ 0 & 2 \end{pmatrix}$
a*b (error!)	undefined!	
−3*B	scalar multiplication	$-3 \cdot A = \begin{pmatrix} -6 & -3 \\ -3 & -3 \end{pmatrix}$
A\B	left division	$A^{-1} \cdot B = \begin{pmatrix} -2 & 0 \\ 3 & 1 \end{pmatrix}$
A/B	right division	$A \cdot B^{-1} = \begin{pmatrix} -\frac{1}{2} & -\frac{3}{2} \\ 0 & 1 \end{pmatrix}$

Operation in MATLAB	Effect	Example
A^2	raising to a power (exponentiation)	$A^2 = \begin{pmatrix} 5 & 3 \\ 3 & 2 \end{pmatrix}$
2^A (error!)	undefined!	
a^2 (error!)	undefined!	

A.2   Arithmetic Operations as Field Operations

The following table shows the contrasting arithmetic operations of MATLAB as field operations.

Operation in MATLAB	Effect	Example
A.+B (error!)	meaningless, since $A + B$ is already term-by-term	
A.−B (error!)	meaningless, since $A - B$ is already term-by-term	
a.+b (error!)	meaningless, since $\vec{a} + \vec{b}$ is already term-by-term	
A.*B	term-by-term product	$A \cdot B = \begin{pmatrix} -2 & 1 \\ 1 & 1 \end{pmatrix}$
a.*b	term-by-term product	$\vec{a}. * \vec{b} = \begin{pmatrix} -2 \\ 1 \end{pmatrix}$
−3.*B (OK!)	scalar multiplication	see Table A.1
A.\B	term-by-term division	$A.\backslash B = \begin{pmatrix} -0.5 & 1 \\ 1 & 1 \end{pmatrix}$
A./B	term-by-term division	$A./B = \begin{pmatrix} -0.5 & 1 \\ 1 & 1 \end{pmatrix}$
A.^2	term-by-term exponentiation	$A^2 = \begin{pmatrix} 4 & 1 \\ 1 & 1 \end{pmatrix}$
2.^A	2 raised to powers given by matrix terms	$2^\Lambda = \begin{pmatrix} 4 & 2 \\ 2 & 2 \end{pmatrix}$
a.^2	term-by-term exponentiation	$a.^2 = \begin{pmatrix} 4 \\ 1 \end{pmatrix}$

Appendix B

ABOUT THE CD-ROM

- Included on the CD-ROM are MATLAB examples found in this text, along with other files needed to run the in-text projects. Also included are simulations, figures from the text, third party software, and files related to topics in MATLAB and Simulink.
- See the "README" files for any specific information/system requirements in the related file folder, but most files will run on Windows 2000 or higher and Linux.
- This book is about MATLAB and SIMULINK software. It is assumed that the reader has access to a computer with MATLAB version 7 or later. The CD-ROM should be readable by any operating system.
- Rubik's Cube R used by permission of *Seven Towns Ltd.* See file "cube2.jpg" on the CD-ROM, under Chapter 3.

NEW RELEASE INFORMATION (R2007b)

M ATLAB and Simulink are software packages themselves. The MathWorks company looks to improve their software by correcting bugs and adding features. A natural question for a software user is, "Do I need to replace the current version that I have?" Often, the answer is "no," unless there is a compelling reason, such as a bug fix on a function used in your simulation. Or perhaps the updated version supports a new feature that you would use. If you want the security of having the latest available software version, The MathWorks provides a "Maintenance Service" for their software, if you pay the maintenance fee.

C.1 BACKWARDS COMPATIBILITY

Software producers are concerned with backwards compatibility, the idea that new versions of software should work with older versions. As a result, it is expected that what worked with version n will also work with version $n + 1$. But the older versions, of course, do not support the new features. For example, using the double-ampersand to mean "AND" did not appear until recently in MATLAB, though other languages commonly use it. A statement such as the following:

```
if ((a == 3) && (b == 0))
    ...
```

may cause a problem, depending on the MATLAB version.

Here is a small program using the double-ampersand to AND two conditions.

```
a = 3;
b = 0;

if ((a == 3) && (b == 0))
    disp('yes');
else
    disp('no');
end
```

We will call it `testversion.m` When we run it with an old version of MATLAB (Release 12), we get the following error.

```
>> testversion
??? Error: File: /home/cscmcw/testversion.m Line: 4
Column: 15
Expected a variable, function, or constant, found "&".
```

Now, we run this same program with a more recent version:

```
>> testversion
yes
```

A way to fix this would be to use the old syntax, i.e., `and(a == 3, b == 0)` We modify the program, now calling it `testversion2`.

```
a = 3;
b = 0;

if (and(a == 3, b == 0))
    disp('yes');
else
    disp('no');
end
```

Now compare the results, first with Release 12.

```
>> testversion2
yes
```

Now we repeat with the new release.

```
>> testversion2
yes
```

The point here is that the newer version of MATLAB understands the older MATLAB code, while the older MATLAB version is not able to handle code made for the newer version. In a situation where multiple version of MATLAB exist, we can specify the code to work with the oldest version, with reasonable certainty that it will function correctly with the newer version, too.

C.2   WHAT IS NEW FOR R2007b

While this book was being written, The MathWorks released new versions of MATLAB (7.5) and Simulink (7.0). This section provides an overview of the new features. Two new products were announced: a new design verifier for Simulink, and a code development package for embedded systems called Link for Analog Devices VisualDSP++®.

The Simulink Design Verifier produces test cases (design verification blocks) to evaluate models. It reveals when elements cannot be reached, as well as gives examples where model properties are violated. It generates a report in HTML, meaning that the user can navigate the report like a webpage. This software can be accessed by the `sldvlib` command at the MATLAB prompt, or by the `simulink` command followed by clicking on the "Simulink Design Verifier."

Processors made by the company Analog Devices, Incorporated, are supported by MATLAB and Simulink. Analog Devices has a product called VisualDSP++ that allows people to program their processors in C/C++, as well as link, debug, and simulate code. These processors are often used in embedded systems, and now MATLAB and Simulink work with VisualDSP++.

One new feature produces C language code from the subset that supports embedded systems, which can be done through the command line interface. Embedded systems are essentially computers, put in the role of a dedicated task. For example, a building-monitoring system may use most of the same hardware as a desktop computer, but have special components and sensors, including a tiny display and keypad for communications with the user. While a desktop computer should have good response, an embedded system must reliably meet deadlines (hard-time constraints) or it is considered a failure. We call this Real-Time computing. Embedded systems have different challenges and concerns than regular computers, so programming languages for them also have special considerations.

The new software from The MathWorks has updates to fix many known bugs, affecting 50 MATLAB toolboxes and 30 Simulink toolboxes. A few other noteworthy changes are listed below.

- The new version supports AVI, MPG, MPEG and WMV video formats, though only on the Microsoft Windows version. The `mmreader` object allows the reading of these file formats.
- Functions in MATLAB will support larger arrays of numbers ($2^{48} - 1$ elements, instead of 2^{31}) on 64-bit computer systems. This should help with video and sound processing.
- MATLAB provides a new exception handling class, called `MException`. It adds capability to the error handling feature, such as allowing the `catch` block to have additional information about the error.
- Custom made toolbars are supported with GUIDE, allowing the user to create GUI toolbars with the mouse instead of programming statements.

Additional information about the new releases can be found on The MathWorks' webpage, `http://www.mathworks.com`.

SOFTWARE INDEX

Index

CPSIA information can be obtained
at www.ICGtesting.com
Printed in the USA
FFOW01n2123290216
21894FF